Design for Assembly

Design for Assembly

Principles and Practice

Alan Redford and Jan Chal

McGRAW-HILL BOOK COMPANY

London · New York · St Louis · San Francisco · Auckland
Bogotá · Caracas · Lisbon · Madrid · Mexico · Milan
Montreal · New Delhi · Panama · Paris · San Juan
São Paulo · Singapore · Sydney · Tokyo · Toronto

Published by
McGRAW-HILL Book Company Europe
Shoppenhangers Road, Maidenhead, Berkshire, SL6 2QL, England
Telephone 0628 23432
Fax 0628 770224

British Library Cataloguing in Publication Data
Redford, A. H.
  Design for Assembly: Principles and
  Practice
  I. Title II. Chal, Jan
  670.42

  ISBN 0-07-707838-1

Library of Congress Cataloging-in-Publication Data
Redford, A. H.
  Design for assembly: principles and practice/Alan Redford and
  Jan Chal.
    p. cm.
  Includes bibliographical references and index.
  ISBN 0-07-707838-1
  1. Assembly-line methods. 2. Production planning. I. Chal, Jan,
  II. Title.
  TS178.4.R43 1994   93-8544
                                                                    CIP

1234   CUP   9765 4

Typeset by BookEns Ltd., Baldock, Herts.
and printed and bound in Great Britain at the University Press, Cambridge.

Contents

Preface

In the past 25 years, there has been a considerable shift in emphasis in manufacturing research. Before then, most effort was directed at primary manufacturing processes until it became clear that the potential for savings resulting from improvements was not covering the cost of the effort. In the high labour cost economies, with more competition arising from the low-cost emerging nations, it was recognized that one virtually untapped source of reduced costs was assembly which, with limited exceptions, was still being carried out in the same way and under the same circumstances as at the turn of the century. It also became apparent that for both manual assembly and for increasing levels of automation, the most effective method of reducing assembly costs was through good product design. In the sixties, books on assembly automation began to appear and these often contained advice on 'design for assembly' (DFA) but this was presented in an unstructured way. It was only about 15 years ago that the first design 'systems' appeared where sets of formal procedures were developed to be used in a systematic way to determine the problems that arise in assembly and to show how these problems might be avoided or marginalized.

In the last 15 years many commercial DFA methodologies have been developed and increasing numbers of companies are taking advantage of the benefits offered by their use. Virtually all DFA procedures address the problems associated with part geometry and inter-part spatial relationships for either existing products or for post-conceptual designs because these give the most immediate and obvious returns on effort. Unfortunately, no attempt has been made to tackle the problems associated with assembly organization or those concerned with conceptual design. Commercial systems are useful but, because they are necessarily constraining to ensure that the user cannot misuse them, they do not offer the user the relationship between the various technologies and the rules that would help the user to appreciate as well as use the rules.

This book attempts to deal with these points. Assembly activities are categorized and design commonalities and differences are discussed in relation to assembly methods, assembly processes, and assembly method and process independency. An attempt is also made to indicate how conceptual design might be accommodated where, at the time; the product and its parts do not have form, shape or size.

It is hoped that on reading this book, product designers will become more familiar with the philosophy of assembly rather than just the mechanics of assembly design and that this will lead to better understanding and subsequently better designs.

PART I

Methods and Processes

CHAPTER 1

Introduction

1.1 Development of manufacturing

Until the late eighteenth century, manufacturing was a craft-based activity in which one person would be responsible for all aspects of manufacturing, at least from the procurement of basic manufacturing materials stage. This method of manufacture has three obvious disadvantages:

1 The manufacture of goods was supply driven rather than demand led, i.e. the time lag between a step-change increase in demand and the ability to meet the demand was so great that it was impossible to meet the increase.

2 The development of new products involving new technologies was inefficient because no common building blocks existed.

3 Manufacturing methods, even after allowing for the constraints imposed by the state-of-the-art of the technology, were inefficient because of the lack of repetitions involved in the work.

The situation changed with the beginning of the industrial revolution when, due to more effective sources of power, wind, water and steam, the first metal-manipulating machines were developed. In metal removal processes, first the drilling and milling machines and then the lathes were produced. In metal deformation processes, the press, the drop forge and primitive extrusion tools were developed. As an aside, it is interesting to note that these primary manufacturing processes remain virtually unchanged and improvements in production have been achieved mainly by improved tool and workpiece materials.

The two main consequences of improved primary manufacturing and, in particular, the development of contouring machines were:

1 The realization of the concept of interchangeability of parts whereby a

set of parts required for a product could be selected at random and assembled to form the product.

2 The increases in production rate that ensued.

This led to a separation between primary manufacturing and assembly which still exists and which will continue to exist in the future. Assembly was still categorized as either fitting or assembling; fitting implies some secondary manufacturing processing to improve or allow product functionality, whereas assembling implies only manipulation of finished parts into meaningful spatial relationships, i.e. generally the skill factor in fitting is significantly higher than that in assembling. In the early days of assembly, fitting, being an independent activity, represented a significant proportion of assembling; this has reduced and currently fitting can be assumed to be negligible.

The next significant manufacturing development that took place resulted from observing the effectiveness of labour in assembly. It was found that the efficiency of assembly improved dramatically when repetitive actions took place, i.e. as well as the learning period which results from exposure to a new set of circumstances and which eventually produces more efficient performance, there is a second learning period resulting from continuous exposure to the same problem which also produces more effective performance but which regresses once other activities are interposed. This led to the concept of limited task assignments.

A further significant development was the pioneering work carried out by F. W. Taylor at the Bethlehem Steel Company in his initial studies of people pushing wheelbarrows, which laid the foundation stones of modern work study. Taylor found that maximizing efforts does not result in maximized efficiency because effort is time dependent and decreases increasingly non-linearly with time as effort approaches maximum. He also found that at some point the reduction of effort required, for increased time, eventually results in optimum performance, i.e. both working too hard or too long results in decreased performance. This is significant in the context of current circumstances; contrary to the views expressed by many, the way to improve manufacturing efficiency generally and assembly efficiency specifically is to reduce assembly content, not to attempt to increase assembly effort.

Work study developed rapidly and the ideas of minimal movement, physical comfort, conducive environment, etc. that are common today were investigated. It is generally accepted that the person most responsible for bringing together the production methodologies which had developed throughout the nineteenth century was Henry Ford. In the early twentieth century he popularized the concept of manual line assembly by his famous initiative which led to the assembly of the flywheel magneto for the Model T on the assembly line.

The concept of manual line assembly is simple. The assembly content is divided, in an acceptable sequence of assembly, into 'equal' portions of work content, the number of portions depending upon the production requirement. When the production requirement is such that the number of portions is small, the cost of the assembly fixture handling system cannot be justified and manual bench assembly, in which every operator assembles the complete product, is appropriate.

The theoretical limit is reached when the production volume is such that all tasks are elemental; an elemental task is one that cannot be further subdivided. The practical limit is usually larger portions of work since the time differences between different types of elemental tasks result in different amounts of effort, and this would be counter-productive.

The portions of work content which have been identified are then carried out sequentially by manual assembly operatives, who are spaced on the assembly line. It quickly became apparent that efficiency was improved by carrying the partial assembly in a fixture and by moving the fixture automatically between assembly operatives (stations). Subsequent work on seating, line height, line speed, lighting, etc. quickly maximized system efficiency. The results Ford obtained were spectacular; the total assembly time for the product was reduced from 30 to 4 minutes and the era of mass production had begun.

Not long after this, it was recognized that many operations being performed by the assembly operatives were simple and that these operations could be carried out automatically by the use of suitable equipment, i.e. operators began to be replaced by automatic workstations. When the way that the operative performed the task was observed and analysed, it was obvious that these workstations could not carry out the task in the same way as an operative. Because of their later importance in the context of modern assembly, these differences and their consequences will be emphasized now.

The methods used by operators for piecepart acquiring and inserting were as follows:

1 One part was picked at a time from a series of containers with parts constrained in pseudo-random orientations.

2 This one part was manipulated (oriented) by the operator into an

orientation which observation and experience had shown to be most appropriate for efficient inserting.

3 The operator then translated the part to the immediate environment of the mating part and, in general, established a contact between the two.

4 Using a combination of senses, predominantly vision and touch, and iterative coordinated movements, the insertion process was initiated.

5 Using one or more motions and much more limited sensing, the insertion process was completed.

The above list has been written in the past tense to be consistent but, of course, the methods used today for manual bench and line assembly are, with limited exceptions, the same.

The alternative methods sometimes used today are:

- Parts are sometimes presented in a specific orientation. This is achieved by either supplying parts in a suitable container (magazine/pallet); by using an automatic small parts feeder which replaces the acquiring and manipulating functions of the operator; or by manufacturing and assembling as linked activities. Magazining/palletizing is considered appropriate in two circumstances, firstly when it is inexpensive and secondly when it is more efficient. By definition, at the point of primary manufacture, part orientation is known and it is only subsequent activities which destroy this orientation. Examples of this are secondary operations such as plating, spraying, degreasing, barrelling, etc.

- Alternatively, when the point of manufacture is remote from the point of assembly, transportation and other costs become significant, and maintaining orientations usually becomes expensive because of magazine or pallet costs and the reduction in packing density associated with the activity. The exceptions to this occur when the magazine is inexpensive (usually throw-away), as is the case for staples, electrical components, or mandatory because of the potential for damage, as is the case for ICPs, glassware, etc.

- Taken together, it is reasonable to assume that for most small bought-out parts, keeping parts in orientation is not practicable and that this method of parts presentation is restricted to large or delicate bought-out parts or in-house manufactured parts.

- Automatic small parts feeders are sometimes used in manual assembly but more often in bench assembly. than line. It is accepted that assembly efficiency will increase but often the improvement does not justify the cost of the equipment.

- Manufacturing at the point of assembly is not often appropriate but it is used for producing items such as shim-washers and springs.

If it is accepted that most parts in typical assemblies are small, predominantly manufactured off-site and usually robust, then, for manual assembly, acquiring parts from containers having unique parts in pseudo-random orientation is nearly always the most effective method.

Once it is accepted that parts are acquired from random orientations, methods of carrying out the subsequent activities have not significantly changed. Power tools and multiple or active fixtures are perhaps used more now, but this is only a matter of degree rather than a basic difference in technology.

Returning to the original argument, it is clear that automatic workstations cannot perform in a way that operatives find effective because the operative uses facilities which cannot be duplicated, would be exorbitantly expensive or would be impractically time-consuming.

One definition of effective assembly is the ability to take a part in an unknown orientation, in a poorly defined location, and to mate it with another part or parts, also in unknown orientation and poorly defined locations, by motions in which the only forces generated result from those needed to grip the part and those that are necessary to mate the part to meet the functionality. With this definition, operatives are good at acquiring and bad at inserting, whereas mechanisms are incapable of either.

If the specification is relaxed such that the part or parts being mated are in a known orientation and a well-defined location, then the operative's effectiveness does not significantly change but the mechanism can now become highly efficient at inserting. If the specification is further relaxed such that the parts being acquired are also in a known orientation, then the operator's effectiveness is marginally improved and the mechanisms can be highly efficient at both acquiring and inserting. This biased argument has been formulated to indicate the basic differences between using people and mechanisms for assembly. Of course, people are extremely effective at both acquiring and inserting parts but, in the context of a somewhat extreme definition of what good insertion requires, people can never achieve good insertion. Alternatively, if the parts' orientations and locations are well constrained using the same definitions, mechanisms can achieve good acquiring and insertion. This is summarized in Table 1.1.

Table 1.1 Capabilities in unconstrained and constrained assembly environments

Assembly condition	Manual assembly	Automated assembly
Poorly defined acquiring location	Good	Very bad
Poorly defined insertion location	Bad	Very bad
Well-defined acquiring location	Good	Good
Well-defined inserting location	Bad	Good

There are two highly significant points to be made from this:

1 People are good at assembly in spite of their lack of certain abilities. People use vision or, for occluded objects, special aptitude to get within range of an assembly task. They then use tactile sensing in coordination with movement to achieve the task.

2 For assembly, trying to emulate people when using mechanisms is not appropriate since the requirement is not to assemble like people, but merely to assemble.

This was recognized in the context of automatic stations. As will be seen in a later chapter, when the flexible assembly system evolved, this principle seems to have been forgotten. The functionality required then of an automatic workstation was:

• a means of converting parts in random orientation to a supply of parts at an appropriate rate in a unique orientation at a well-defined location,

• a means of acquiring the part,

• a means of transporting the part,

• a means of inserting the part in a unique but often different orientation from that of acquiring in a second well-defined location.

The mechanisms used to meet the requirements, and further useful constraints, will be dealt with in Chapter 4. At this point it is sufficient to say that it is now possible and economic to use automatic workstations.

At the time that the first manual assembly lines were used, labour for operating the lines was readily available and the monetary rewards were significant compared with other forms of employment available to the semi-skilled and unskilled worker. This tended to compensate very much for the repetitive boring work involved, and there were no problems with recruiting and keeping a viable work-force. Progressively, since that time, with a continuing improvement in living standards within countries with a strong manufacturing base, labour's perception of what is acceptable working practice has shifted and it is now increasingly difficult to recruit operatives to work on assembly lines. It is found that, in this kind of activity, absenteeism is high, there is often a very high rate of labour turnover, pride in the work is almost non-existent and job satisfaction is low. The result of this is a loss of ability to control manufacture, and poor quality products.

Because of this and the changing nature of the work, companies are turning to different methods of manual assembly which combine the interest of bench assembly and the efficiency of line assembly. These hybrid manual assembly methods will be discussed later in Chapter 5.

The introduction of manual and automatic assembly lines marked the beginning of the mass production era, which by definition was the beginning of consumer-driven society. At this time, simple market-force rules applied. As the cost of products reduced, the demand for them increased, which in time led to lower costs and more demand, etc. To the consumers of the day, goods became available which had previously only been affordable by the wealthy few. The novelty effect of this was important; for the majority, the quality of life was poor, and almost anything that improved this was welcome. This led to a society for which significant improvements in living standards were achieved but for which there was severely restricted choice.

It could have been expected that this period of development in society would have been relatively short but this was not the case, primarily because of two major wars with a long and deep depression between. As a result of this, and because of the time needed after 1945 to change a wartime economy into a peacetime economy, with limited exceptions the products available in the early fifties were not significantly different in technology and choice to those available very much earlier in the century.

Since this time, three major developments have taken place which have completely revolutionized the manufacturing industry. Firstly, on the production side, the computer age has led to improvement in the control of manufacturing systems and processes; simultaneously this has led to very many more consumer products becoming available. Secondly, in the last 40 years, the development of the plastics industry has led to their widespread use, particularly in consumer products, which has also resulted in more and lower-cost products. Thirdly, as a result of reduced real costs and more products, marketing has exploited these advantageous conditions to educate the consumer into demanding not just different products but more variety of basically the same product. This has led to a manufacturing nightmare of more and more products with more and more variants, which in turn has led to very much smaller total production volumes of any particular variant and, equally importantly, the production of much smaller batches. This has caused significant problems both in the equipment for, and the organization of, manufacture.

Fortunately, the equipment for primary manufacture has kept apace with market requirements primarily because of the continuous and continuing reductions in cost of computing power. Simultaneously, the organization of manufacturing has also progressed substantially and the complex interactions between the various aspects of production are well understood. It could be argued that the above is the cause rather than the effect, i.e. it was only when equipment became available to challenge the concept of large batch production that marketing was able to exploit this; if this is true then it is probably the biggest own-goal ever inflicted by a major commercial sector on itself.

If the equipment is available and the technology in place and the increased costs associated with this for piecepart manufacture are minimal, what is the disadvantage of giving the consumer more choice, since it could be claimed that this improves the 'quality' of life? The big disadvantage is that for the products being considered, a large portion of manufacturing cost is associated with assembly, and assembly equipment has not yet been developed which can economically accommodate small batch production of products in small total production volumes. Automatic assembly machines cannot recover the investment on small volumes and, even if they could, they cannot be reconfigured to suit different products; manual assembly, while clearly capable of the tasks required, is very much more expensive. This problem has been resolved very simply. In any manufacturing situation where the technology is available but the costs are too high, either methods of reducing the costs are sought or alternative technologies are developed. In this case, a ready-made low-cost solution was obvious. Since it was the cost of labour rather than the technology associated with labour which was the problem, move the work to where labour costs are low.

Maybe it was felt that automation would quickly be able to accommodate the new manufacturing conditions and that this 'new' manufacturing base would only be transient; this has proved to be an extremely optimistic assessment. The supposed saviour in the automation of assembly for the new market-led manufacturing industry was to be the industrial robot. However, it is now 25 years since the first commercial robots became available and they have not yet made any kind of impact on assembly technology. The reasons for this and the interactions between product change and equipment capability will be expanded in Chapters 4 and 5. Because the prominent manufacturing nations of the last generation have effectively exported jobs by changing the specification for manufacturing, what are the alternatives available?

Withdraw from manufacturing
This is the simplest alternative and in Britain particularly, it is not too long ago that this was seriously considered as a viable option. Thankfully this possibility has lost favour because, presumably, those with some ability to control the agenda have made the obvious correlation between wealth and manufacturing; it is no accident that with limited exceptions such as control of strategic natural resources, nations that are good at manufacturing are wealthy, i.e. many activities move money round within a society but manufacturing, above any other, has the potential for creating 'new' money.

Concentrate on 'high-technology' production
The necessary technology is not yet usually available to low labour-cost economies and the added value is very high. This has its attractions and champions but does not stand up to critical inspections when considered even superficially. Individual people are by far the biggest consumers of manufactured goods and there is nothing to suggest that this will not always be the case. Relatively trivial products dominate manufacturing and a viable industry has to address the problem of manufacturing these types of goods. It is no surprise that many of the most successful patents have been for the manufacture of products which are relevant to people, e.g. cats'-eyes, supermarket trolleys, home work-benches; even more advanced products such as Xerox machines and computers are still people-dominated. Further evidence supporting this hypothesis is that when we run a balance of payments deficit this is always linked to consumer buying power, and there is a call for cut-backs in consumer spending to halt the flow of incoming goods.

Embrace the problem of assembly automation
Take the initiative and try to automate the assembly of low-cost, low value-added products. While, at first sight, the obstacles to be overcome appear daunting, there are other effects that come into play which are worthy of note.

Firstly, it has always been the case that for situations in which people compete with machines, the machines become relatively less expensive with the passage of time; this is because the machines become more effective, they are manufactured in bigger volumes, which reduces costs and they are eventually designed for economic manufacture. Thus, with no other changes, it can be concluded that at some point in the future, automated assembly appropriate to market requirements will be cost-effective.

Secondly, as wealth moves between societies, the new owners of wealth expect the rewards of wealth and a general levelling takes place; this accelerates wage inflation at the bottom and reduces the time to the break-even point between people and machines. A complementary argument is that products only have value if they are wanted and affordable—particularly affordable. As more products become afford-able there is greater demand and whereas labour costs are unaffected by demand, equipment costs per unit of demand reduce.

A third argument is that continuing increases in wealth in a minority of society with relatively decreasing wealth for the majority is not sustainable; in demand-led economic systems, the analogy of having a gold brick on a desert island is appropriate. A rider to this is that, even without 'market forces' in all societies, wealth and poverty can only diverge for a limited period after which there is often a step-change convergence.

What are the consequences of letting events happen by not tackling the problem of widespread assembly automation? Others will, and eventually they will start to show results on investment. Manufacturing costs will increase relatively, primarily because of assembly, and market share will be lost. The emphasis would have to move to high value-added with even smaller production volumes and batch sizes. This will inevitably require less manufacturing capacity and will result in reduced wealth creation; the pendulum will swing and wealth will probably reduce for several generations regardless of how individual nations act. The outcome worldwide is predictable; as national or international wealth increases, individual wealth follows and labour-based manufacturing economics become less competitive. This will be counteracted by less reliance on labour, which will continue a trend that started 300 years ago. Eventually, as happened to agriculture, in manufacturing societies labour will become insignificant in manufacturing. It is difficult to predict the consequences of this and the outcome is not pertinent to this text. What is under consideration is, given that manufacturing is important and that manufacturing of trivial products is necessary, how can manufacturing economies be improved?

1.2 Improved manufacturing economics

If manufacturing economics are to be enhanced by reducing the reliance on labour, how might this be achieved? The following considers some of the alternatives.

Less customer choice

Attempts could be made to reverse the trends of the past few decades by 'educating' the customer to require less choice. This would undoubtedly produce the required result but it is virtually inconceivable that it is a valid argument. There is much evidence that the old saying 'what you have never had, you never miss' is true, but the converse is also true. Once consumers are exposed to an environment in which they perceive there is an 'improvement', it is very difficult to change the perception, particularly to a situation where the 'benefits' are less obvious and more particularly when it requires a return to a discarded purchasing ethic.

A good example of this is the car. Not only do the consumers demand endless models and variants, they also persevere with using a car in situations where often both the quantity and quality of life are impaired. Henry Ford had a sound economic argument when he built black Model Ts but the company nearly became insolvent when other manufacturers appealed to the customers' innate individualism.

Improved piecepart manufacturing technology

This could take one of two basic forms:

1 Manufacturing equipment could be improved and/or new manufacturing equipment could be developed.

2 New materials could be developed for the manufacture of either products or the equipment used to manufacture the products.

Certainly, over the last 100 years, such improvements have made a contribution to reduced manufacturing costs but because of the advances already made and because the rate of advance is naturally reducing (the law of diminishing returns), it is difficult to see significant further benefits.

Most new equipment is only a reconfiguration of existing primitives and it really should be categorized as improvements. Genuine new technology equipment does appear but is often highly specialized and makes very little impact on manufacturing in total.

In the past, the development of new materials has perhaps had the most significant impact on manufacturing costs. It is often argued that the industrial revolution and the subsequent industrialization of the world have been almost exclusively the result of improved materials and improved methods of measurement. It is difficult to imagine how modern manufacturing equipment could have developed effectively without the cutting and forming tools and the workpiece materials that we now have. It is equally difficult to see, as mentioned previously, how modern products could have evolved without the work of polymer scientists. In the future, there is no doubt that the materials' scientists will have a large role to play in reducing manufacturing costs.

From the arguments it is clear that the effectiveness of improved piecepart manufacture in reducing manufacturing costs has limited potential and that other, more cost-effective possibilities need to be examined.

Improved design for manufacture

Product design for piecepart manufacturing—where the main aims are designing so that less expensive, more appropriate or a lesser quantity of equipment can be used—is well advanced. Most designers know a lot about their products and, perhaps more importantly, about manufacturing equipment, its specification and limitation, and materials. The advent of CAD (computer-aided design) and its subsequent development has reduced design time and improved design quality. This reduces design costs and lead time and substantial savings have been made, but it is also well advanced by now and the potential for improvements is limited.

Improved control of manufacturing systems

In the last two decades, perhaps the biggest contributor to improved manufacturing efficiency and hence reduced costs has been improvements in the organization of manufacture, particularly to suit the changing and more demanding manufacturing environment. What has changed, and is changing, in this context is the variety of work, with the subsequent conflicting demands that variety requires. Manufacturing unique products on a continuous or semi-continuous basis has no conflicts; raw materials enter the system at one end, their progress through the system is hardly perturbed by future events, and they leave as products at the far end. Meeting demand within the system capacity is simple, work-in-progress is minimal and it is not difficult to estimate minimum stock levels of both raw materials and finished goods, given a demand profile. The benefits of effective system control can be validated by reference to Japanese industry; it is clear that this most successful manufacturing country has been the originator and most successful developer of effective systems. The Japanese have recognized that the work-in-progress resulting from ineffective organization of manufacture is not only a liability which is accruing debt and to which value has not been added, but a potential asset for which the transition from liability to asset is being delayed.

It is important to note that, despite what the popular press might say to the contrary, this has been achieved with no significant advances in manufacturing technology. Indeed, it can be argued that the engine behind Japanese manufacturing, the SMEs (small manufacturing enterprises) supplying components, are certainly no better and sometimes worse than their counterparts in other countries.

Advances in the understanding of manufacturing systems have developed rapidly in the past 20 years. While it is more than likely that enhanced and alternative systems will evolve, the control strategies will probably be sufficient to accommodate these changes and the potential for significant improvements will not be a result of more effective control but of more effective systems.

Improved manufacturing systems

As mentioned above, it is inevitable that changes in the manufacturing environment resulting from changes in demand will cause an evolution of different manufacturing systems which might be more, or less, efficient than those that already exist. However, when addressing the problem of what can be done now to improve what exists now, then by definition, improved manufacturing systems responding to different demands do not form part of the argument.

Improved assembly

After 200 years of having the ability to mass manufacture pieceparts which are interchangeable, and after 100 years of being able to show the economic benefits of improved assembly efficiency, at last assembly is now quite near to the top of the agenda at meetings that discuss and determine manufacturing strategy. Why has this taken so long when it is well known that, for most of the types of product under consideration in this book, typically 50 per cent of the manufacturing cost is a result of the cost of assembly? There are many reasons for this; below are several of the more important ones.

- To manufacture pieceparts requires power. The removal of, plastically deforming of, melting of, metals etc. cannot be done to any marked extent by people using only crude inexpensive tools. It is incredibly difficult to make ball-bearings using a cold chisel and not much easier using a hacksaw and file, i.e. when it comes to priorities it is sensible to tackle the most difficult, or restricting, problems first and since people with fairly crude, unsophisticated tools can assemble things, why put effort into developing assembly when there is a greater need to develop methods of manufacturing parts?

- Until there is the ability to make pieceparts there is no assembly problem. Obviously it has made sense to initially concentrate on primary manufacture.

- People are good at assembly. Of all the species on earth it is claimed that the aptitude which distinguishes humans from all the others is the ability to fabricate and use tools and hence fashion artefacts which have more than one part; this implies an ability to visualize geometric and spatial relationships, a prerequisite for the ability to assemble.

- Everybody thinks they are experts at assembly. Because of the instinctive ability of people to put things together, assembly is often not even considered to be a discipline and hence there is no need to formulate 'rules' for assembly which can then be used to improve ability and efficiency.

- People are familiar with assembly. Because we are instinctive assemblers and take pleasure from that activity we are subjected to an assembly environment from an early age. It is not surprising that most successful toys are based on the aptitude for assembly, e.g. Meccano, Lego, jigsaw puzzles, etc. As the old saying goes, 'familiarity breeds contempt' and people have their own ideas about assembly. Picture a typical building committee meeting; the plans for a new £2 million clean-room pass on the nod, yet new toilets at £2000 demand endless arguments. Very few people know a lot about clean-rooms but everybody knows what a toilet is. Similarly, because people know about assembly they have prejudices and will argue they are 'right'.

- Assembly is inexpensive. Capital investment for manual bench assembly is minimal and this implies that, for this form of assembly, costs are almost all labour costs. In the end, if labour costs are zero, what is the incentive to improve assembly efficiency? As mentioned previously, the rate of increase of labour costs has been greater than the rate of increase of equipment costs for many years, but it is only recently that the labour costs in the major manufacturing countries have reached a level at which widespread investment in assembly automation has become necessary. It is not surprising that less well-developed manufacturing countries, to be competitive, have chosen to concentrate on high labour content work. In reality, this means assembling consumer products.

1.3 Improvements in technology

The reason why the problems of assembly have not been considered in the past to be of high priority have been outlined above. In the now-changed circumstances, it has been argued that to compete effectively in the future will require much higher levels of automation, but this presupposes that improvements in technology will lead to improvements in efficiency; the potential for this will now be discussed.

Improved assembly technology

Later in this chapter, various methods of assembly will be described and it will be evident that of the three major classifications of assembly methods, manual assembly, dedicated assembly and flexible assembly, only the latter has development potential. Manual assembly methods, whether bench, line or some hybrid containing elements of each, are both well established and well researched. It is not conceivable that significant improvements in the basic methods can be achieved. Indeed, virtually all recent improvements have resulted from either better control of the manufacturing system, particularly materials flow or improved job satisfaction, with the result of less labour turnover, less absenteeism, improved quality of product, etc. The improvements have not been due to improvements in the basic technology.

Dedicated assembly systems have been in operation for probably 80 years and it is difficult to see very much change in these. Indeed, attempts to modernize using, in particular, pneumatics have usually resulted in equipment inferior to that using the original concept of a cam-operated mechanism with spring-in, cam-out actuation. The only real improvement that has taken place has been the introduction of PLCs, (programmable logic controllers) although the main benefits of these are lost on cam-operated machines. What is sure is that when dedicated assembly machines operate under suitable conditions there is no other assembly method which can remotely compete in terms of cost-effectiveness.

Flexible assembly systems have been in use for 20 years and, with limited exceptions, have been an unmitigated disaster. The exceptions have been flexible line-assembly systems producing large volumes of limited variants; in the context of the full spectrum of assembly activity, this is one relatively insignificant element of the total requirement. It follows, therefore, that there is no equipment available which can compete economically with manual assembly for small-batch, small-volume production of trivial products (sub-assemblies) and that, by definition, the potential for the development of suitable equipment is high.

Improved design for assembly

Many of the arguments quoted previously as to why assembly has always been the poor relation in manufacturing are as equally valid for product design for assembly as they are for lack of activity in assembly technology. To these can be added other observations which are specifically pertinent to design.

1 Designers have many tasks to perform. They are responsible for conceptual design, design for functionality, design for manufacturability, design for appearance, design for reliability and design for assembly. With the exceptions of design for manufacturability and assembly, all other design activities are apparent to the customer and if demand is to be created, these design activities must take place; this tends to lead the designer to prioritize these.

2 Because manufacturing equipment has known capability and most designers have been exposed to the equipment, they know about the equipment and its capability and they design accordingly. For example, it is not coincidental that most parts produced by metal removal are normally axi-symmetric or rectangular prismatic; lathes, grinders and milling machines, etc. produce these shapes.

3 With limited time and an ever-increasing demand for shorter lead times, if corners are to be cut they will be cut from what is considered to be the least important activity, i.e. assembly.

4 Because people are good at assembly, the design criteria for assembly are those that consume the least design time. Often the result is the criterion: 'if it's possible, it's acceptable'. This has often gone unchallenged because people are good assemblers and they can accommodate difficult assembly situations.

All this has inhibited the development of design for assembly and it is only in the past decade that design-for-assembly methods have been developed. The potential for benefit is high; it has been found by the users of design-for-assembly techniques that, typically, 20 to 30 per cent of assembly cost can be eliminated when the design is compared with a 'traditional design'. There is also often a saving of 10 to 15 per cent on manufacturing cost as a result of design for assembly; this activity gives the most significant saving that can be made. Importantly, it is also an inexpensive activity.

The ramifications of this are highly significant. It has already been stated that the assembly of trivial products dominates manufacturing and, in particular, consumer products. These activities are highly competitive and profit margins are often a very small percentage of costs, i.e. a 10 per cent decrease in manufacturing cost often represents a very high percentage increase in profit. Conversely, this kind of saving can convert an unsuccessful quotation into a successful one.

Figure 1.1 shows a schematic of the relationship between implementation and time for the various manufacturing activities described previously. The figure is not meant to be accurate but is included to give the reader a feel for where effort would be most meaningful. There are all kinds of inferences about proportion of cost attributable to activities, cost of activities, etc. which have conveniently been ignored to overstate the message that 'design for assembly' and flexible assembly are the most suitable manufacturing activities for implementation and development respectively.

1.4 Objectives of this book

The objectives of this text are not just to list alternative design-for-assembly methodologies and to comment on their effectiveness; this will be done in Chapters 6, 7, 8 and 9 and may be considered to be a catalogue of methods from which a particular method or methods can be chosen to

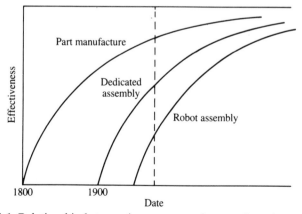

Figure 1.1 Relationship between improvements in manufacturing and time

suit specific circumstances. The main objective is to describe the various assembly technologies and to link the constraints of the technologies to the rules that have been formulated for design for assembly. It is hoped that this will increase the designer's perception of what conditions the rules for design for assembly, and that this will improve the designer's ability to use these rules.

A secondary, and very important objective, is to look, fairly simplistically, at possible extensions to design for assembly, particularly for conceptual design. All current systems effectively start with the assumption that the conceptual work has already taken place, and while this is still very important in reducing manufacturing costs, many of the costs of manufacturing are predetermined once the conceptual design is finished. Chapter 10 will examine the possibilities for design for assembly at the conceptual stage as well as outlining the many factors that are involved in good design for assembly.

CHAPTER 2

Assembly Methods and Processes

2.1 Introduction

As in all good design methodologies, in assembly it is important to decide which factors have to be considered independently of assembly methods or processes, which factors depend on assembly methods, which factors are common to all assembly processes and which are specific to particular assembly processes. With these established, it is necessary to determine the significance of each of these elements to the overall objectives, so that, given conflict, priorities can be established to determine what, on balance, is the best course of action. It is suggested that the first objective of any manufacturing design activity is to achieve functionality and the next objective is to reduce costs. In this context, functionality is considered in its widest sense to include reliability, quality, etc.; this cannot be compromised. It is assumed, therefore, that the purpose of good design for assembly ultimately relates to cost, which involves deciding on the most appropriate process and then designing to accommodate the strengths and weaknesses of the process.

In the following sections the various factors outlined above will be considered in turn and their influence on cost will be established. Firstly, however, it is necessary to indicate the significant different methods and processes and to limit these to what is meaningful in terms of product design.

The two basic classes of assembly processes are those performed by people (manual) and those performed by mechanisms (automated). Manual assembly has many forms, from bench to line; for the former one person is responsible for the assembly of the complete unit and for the latter each person is responsible for the assembly of only a small portion of the complete unit. Each method and its numerous variants have merit under appropriate circumstances but, in the interest of the current context, do they need different design rules because of the nature of the method? The answer to this is yes. To consider the independent factors and the manual assembly-specific process factors is not sufficient—the different methods of assembly require further consideration.

Automated assembly also divides conveniently into two categories, dedicated (hard, automatic, special purpose) and flexible (robotic, general purpose). What, if anything, is different about these? Automated assembly invariably involves the progressive assembly of a unit in which each element of the assembly system is responsible for only one assembly activity (ignoring multiples of the same activity) and, more importantly, is only *capable* of that activity. Implicit in this is that the only flexibility in these systems is the flexibility resulting from the presence or absence of a specific activity. In flexible assembly, however, each element of the assembly system is *always* responsible for more than one assembly activity and has to have the capability for more than one activity. For the former, good design has to address and respond to the strengths and weaknesses of dedicated equipment (parts feeders and automatic workheads). For the latter, the flexibility of the equipment will usually result in some aspects of design being less important but others being more important. The argument suggests that design rules will be different and that dedicated assembly and flexible assembly need some different design-for-assembly rules.

Dedicated assembly equipment also exists in a wide variety of different forms. There are rotary and in-line machines, there are indexing (synchronized) and free transfer (asynchronous), and there are stopping machines and memory pin machines. Adding to the complexity, for the memory pin machines there needs to be a rework or scrapping policy. What is different about all these? Certainly indexing and free transfer, as will be seen, require different design strategies as do different rework policies, but these are separate considerations. The assembly process itself using automatic parts feeders and automatic workheads is identical and, as such, the design rules for one are appropriate to the others.

For flexible assembly, the possibilities for equipment are even more bewildering. In addition to having both bench (single station) and all the options of line alternatives with various control and rework strategies, there are other options such as error recovering, gripper changing, etc. which can influence product design. However, similar arguments can be used to those for dedicated equipment. Regardless of the method of assembly, the flexible assembly process itself uses equipment that is always responsible for more than one assembly activity and needs its own set of different design rules as outlined earlier.

Some important factors have emerged:

1 There are factors which are independent of the assembly methods and process but which require good design for assembly.

2 There are product design factors which are independent of the assembly process but which depend on the assembly methods. Examples of these are bench and line assembly, synchronous and asynchronous systems, stopping and memory pin machines, rework and scrap strategies.

3 There are product design factors which are dependent on the assembly process where the three basic processes are:
(a) manual assembly,
(b) assembly with equipment capable of only one activity,
(c) assembly with equipment capable of more than one activity.

For these, the product design needs to consider:
(a) factors common to the three basic processes,
(b) factors specific to the process.

Later, all of these will be linked to product design.

2.2 Factors determining assembly method and process

Before correlating factors with processes, the factors need to be identified and the methods and processes categorized. Significant factors are few and can ultimately be resolved into three factors: how many, how many variants and how much.

Clearly, some of these factors cross-relate but, simplistically, they show how assembly methods and processes are identified. For example, if the two basic assembly methods are bench (single station) and line, then at some stage the total number of parts becomes such that line assembly becomes more effective than bench assembly (as a result of the difficulty of managing parts). Similarly, at some rate of production, line assembly becomes more attractive than bench (due to less duplication of equipment). In the same way, if the three basic assembly processes are manual, dedicated and flexible, as the frequency of requirement of a particular part reduces, at some stage manual assembly becomes the only viable option; i.e. the cost of manual assembly is nominally independent of what it is, whereas the cost of flexible or dedicated equipment increases with reduced utilization (more variants). Lastly, if labour is inexpensive, manual assembly must be the most attractive (if labour is free who needs automated assembly!).

These observations can be further qualified, using realistic figures for volumes, variants and cost gains:

• Only ten per cent of products have volumes suitable for line assembly.

• Only ten per cent of products have few enough variants for dedicated assembly.

• Ninety-five per cent of products are unsuitable for flexible assembly because of low labour costs.

It becomes apparent from this that most assembly is manual and, while it is common to extol the virtues of product design for other forms of assembly, it *currently* has an insignificant effect on manufacturing. However, for a particular industry or a particular factory which uses either dedicated or flexible assembly equipment, product design for these processes is very important; these issues will be addressed in Chapter 4.

The objectives of assembly can be listed as:

• The bringing together of a number of objects and placing them in a meaningful spatial and physical relationship relative to each other.

• The sequencing of these tasks within the constraints imposed by the design such that they cause minimum disruption to the end objective.

• This could be defined as minimizing effort (cost) until the probability of success is most appropriate, applying effort in a way that minimizes cost in the event of failure or applying effort in such a way that failure can be converted, at least, into a partial success.

2.3 Success and failure

To understand the last point mentioned above needs 'success' and 'failure' to be defined. Success can simply be defined as the ability to perform the predefined tasks without any adverse perturbation. Partial success, or partial failure, can be defined as achieving part of the task whereby a judgement needs to be made as to whether, at this point, the 'product' has intrinsic value at the point of failure or whether the failure is economically recoverable such that the task can be completed. Failure is when the results of effort are lost because the cost of recovery is greater than the value at the point of failure.

2.4 Causes of failure

It is also necessary to understand what usually causes failure. In assembly terms, failure results from an inability to place parts in the necessary spatial relationships. This can be caused by the mechanisms of assembly or by the geometry of the parts. Realistically, the former, for any method of assembly, has a negligible effect when considering failure overall. In manual assembly with only simple reliable assembly aids usually used, the reliability of the 'equipment' is obviously high. In dedicated assembly, while mechanisms, actuators and control equipment do occasionally fail, in the context of total failure time these failures are insignificant. Even in flexible assembly where, because of its short history, ultra-reliable manipulators and end effectors are often not available, failure of the equipment is still insignificant.

Failures, therefore, are either as a result of attempting to assemble faulty parts or attempting to assemble misaligned parts. It can be argued that with good control of manufacturing processes, adequate inspection and good housekeeping, faulty parts should be a rarity. The argument is not this simple; in typical sub-assemblies many of the parts, particularly fasteners, are manufactured by specialists with the emphasis usually on using production methods that give minimum costs. Unfortunately these methods also result in less than perfect quality. Improving quality is easy; intrinsically more 'reliable' production methods can be used or the imperfect parts using existing methods can be inspected. Both these alternatives result in increased costs and it is clear that perfect quality implies infinite cost, i.e. even the most reliable manufacturing methods and the most exhaustive inspection cannot guarantee perfect quality. As a consequence of this, there is a level of quality which provides the best compromise between cost of parts and cost of assembly, and this will be discussed in Chapter 4.

Misalignment of parts may be caused by:

- faulty grasping of the part to be inserted,

- faulty positioning of the part to be inserted,

- faulty positioning of the mating part.

For manual assembly, while insertion of parts can sometimes be difficult, misalignments as described above never result in failure. Mechanisms, however, are not precise positioners and, whether it is grasping or inserting, there is a tolerance on the positional accuracy of the various elements in the process which, when compounded, can be significant. If no attention is paid to this, faulty misalignment will be common and will often incur prohibitive costs. The solution, of course, relies on good part design, and this will be dealt with in Chapter 5 where assembly methods are discussed.

For manual assembly, parts that are out of tolerance are the only significant type of fault resulting in failure and the overall effect of this on assembly efficiency is low, i.e. assuming that manual assembly operatives assemble parts perfectly is a reasonable assumption. For this form of assembly, the aim of good product design is then only to reduce the time it takes to carry out the operations.

For automated assembly there are three other types of faulty part; the most common of these is the absence of a part (to be grasped). Grasping of a wrong but similar part (bad housekeeping), or the grasping of other spurious objects, occur less frequently.

Clearly, since any assembly task involves bringing at least two parts together, the combined faults of the parts give the overall quality. The implications of the fault not being associated with the parts being added will be dealt with in Chapter 4 where error recovery is discussed.

In this chapter, an attempt has been made to set the scene as to what are the important methods and processes of assembly and what ultimately conditions these. Success and failure in assembly terms have been defined and the concept of limited success introduced; it is suggested that a fundamental aspect of assembly philosophy is that the attempt to assemble perfectly, while laudable, is invariably impractical in that it is either too expensive, too slow or both. As an alternative, it is suggested that embodying into methods and processes the facility for either minimizing the effects of failure or converting potential failure into partial success is the way forward, particularly for automated assembly. As will be shown in the next three chapters, the relative frequencies of success and failure in assembly are very much conditioned by the quality of design for assembly.

PART II

Product Design Factors

CHAPTER 3

Product Design Factors Independent of Assembly Methods and Processes

3.1 Introduction

Assembly methods and processes are affected by product design. However, there are some aspects of good design which do not depend on the assembly method or the assembly process but which have a major role in design. Since these aspects of design are, by definition, common to, and benefit all assembly, it is imperative that not only are they considered but they are considered at an early stage of product design.

3.2 Number of operations in the product

It is common in design for assembly to suggest design methods that reduce the number of parts in product. This, however, is not in its own right the most appropriate strategy, since the element determining the assembly time and cost is the number of operations required. All parts have at least three operations associated with them—grasping, moving and inserting—and for many parts these are the only operations. By definition, therefore, reducing the number of parts must reduce the number of operations and must result in improved assembly. However, some parts have more operations than the three suggested above and some operations are completely independent of parts; it is important that these extra operations are also considered.

The benefits of reducing the number of operations are:

- reduction in assembly time or cost depending on the assembly methods and processes,

- the potential for less failures and hence higher, less expensive production rates,

- the potential for less in-process inspection,

- higher product reliability,

- lower manufacturing costs,

- faster implementation, and

- the practical requirement of being able to assemble the product.

Most good commercial design-for-assembly systems address the problem of reducing the number of parts. The criteria that apply to the need for any individual part in an assembly are few and well established and can be identified as:

- the necessity for using different materials,

- the requirements for relative motion, and

- the requirement for the product to be capable of being assembled.

Because reducing the number of parts to a minimum consistent with the specification for the product is essentially easy, why do many products have more parts than is strictly necessary? There are two main reasons for this:

1 The single part that has the functionality of the two parts it replaces may have an increased manufacturing and assembly cost which is greater than the combined costs of the originals. This is not inevitable, and experience has indicated that the total manufacturing cost is less for a design which significantly reduces the part count but usually not to the extent that it is minimized.

2 Most designers rely heavily on experience and, in the process, examine the design of 'similar' products. This makes sense but a problem that results is that this reliance tends to ignore improvements in manufacturing methods and materials.

There are three aspects of part reduction that should be considered:

1 Two parts are already of the same material and could be combined by using new manufacturing processes.

2 The two parts are of the same material that could be combined if a different material with an appropriate manufacturing process were used.

3 The two parts are of different material that could be combined if a

different material with an appropriate manufacturing process were used.

The requirements for relative motion are usually self-evident and if it is assumed that the funtionality is being met by good mechanical design it is not common for the number of parts to be reduced by the straight application of the criterion. One 'grey' area in this is where springs are involved—these usually have no motion at one end but relative motion at the other. For most springs, the mechanical specification criterion applies as well and then there is no conflict. For non-metallic springs, however, the spring material can be the same as that at the end and there are many examples when the spring could be integral.

The assemblability criterion is obvious and is only a constraint applied to the other two criteria.

Methods of reducing the number of 'extra' operations required are less easy to define and first it is necessary to define these extra operations. They fall into two categories, those associated with parts and those associated with the assembly processes; the latter will be discussed in Chapter 4. It is perhaps useful to give examples of these two types of operation. Extra operations associated with parts are clearly those for which the extra operation cannot be carried out at the same time as the basic operations. Good examples of this are evident in riveting and are illustrated in Fig. 3.1. Clearly, because the rivet can only be 'set' when all the parts it secures are in place, if the rivet gives a location feature which is useful for placing the parts to be secured, then it is sensible to place the rivet first, and the setting operation cannot occur at the same time as the rivet is placed. In most cases, there is no solution to this particular problem; the operations are usually associated with fasteners requiring plastic deformation which unfortunately cannot often be replaced by alternative, more economic, fastening methods.

An example of extra operations associated with a process is gripper changing in flexible assembly, where the operations are: acquire a gripper, acquire a part, insert a part and remove the gripper. Clearly, any

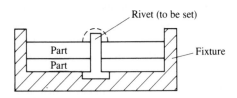

Figure 3.1 An example of an operation which cannot be performed when the part is inserted

design activity which eliminates or reduces the need for gripper changing must reduce the number of operations and increase the efficiency of assembly.

Care must be taken when attempting to reduce the number of operations that there is a genuine reduction in the number. Substituting one operation for a 'less expensive' one, while useful, is only a second-order effect compared with eliminating the operation. Reducing the number of repeats of an operation, say nut-running, is quite useful but often has only a small effect on assembly cost. Considering alternative fastening methods, say adhesive dispensing or welding, is fraught with difficulties, particularly if it is a new technology to the user; it is often also expensive.

3.3 Assembly precedence

One of the most useful tools available for choosing the assembly sequence of a particular product to maximize the performance for a particular assembly method is the precedence diagram. The 'best' sequence is conditioned by many—often conflicting—requirements, some of which depend on factors independent of methods and processes, some of which depend on methods and others on processes. In this section attention will only be given to the former; the others will be considered in the appropriate chapters.

Precedence diagrams

Before alternative assembly sequences can be considered it is necessary to indicate how a precedence diagram is constructed by using a product as an example.

Figure 3.2 shows an exploded view of a simple domestic electrical product, a three-pin power plug. The precedence diagram shown in Fig. 3.3 is drawn by examining how the product may be assembled part (operation) by part (operation). It is assumed that assembly will be in one or more assembly fixtures. The initial judgement is: can or must the product be split into meaningful sub-assemblies? A sub-assembly (S/A) is defined as a group of parts in which *all* the parts are secured and which, from an assembly viewpoint, can be considered to be one part. This product has four sub-assemblies, the base S/A, the earth pin S/A, the neutral pin S/A and the fuse clip S/A. The latter three all contain a terminal screw which can be economically inserted at the point of manufacture. The base S/A could be considered as a part but it is assumed for this study that it is part of the assembly.

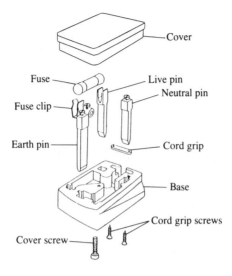

Figure 3.2 An exploded view of a domestic power plug

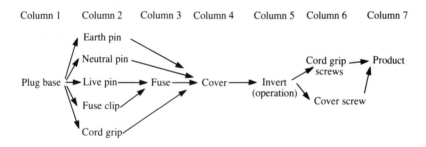

Figure 3.3 Precedence diagram for the power plug of Fig. 3.2

The next judgement is to what is the first (base) part of the product to be assembled? There is often not a unique first part, but in most cases there is an obvious one.

In this case the first part is considered to be the plug base; this is placed in column 1 of the precedence diagram. When the base is added to the fixture, in order to construct the precedence diagram the question 'what *can* be assembled next?' needs to be answered. In this case it is any of the cord grip, the earth pin, the neutral pin, the live pin or the fuse clip; these all are shown in column 2 of the precedence diagram. The fuse can only be added after the live pin and the fuse clip; this goes in column 3 and is linked to the two parts it 'needs'. The next part has to be the cover (column 4) at which point an operation is required (turning the S/A over, column 5). After this, the cord grip screws and the cover screw can be assembled (column 6) and the product is fully assembled (column 7).

From the precedence diagram, all the possible sequences of assembly can be determined and sensible choices can be considered. Good product design should reflect good precedence and to illustrate this, two extreme precedence diagrams are considered.

Figure 3.4 shows the perfect precedence diagram. Here, any part can be assembled at any point in the sequence and there is total choice. Figure 3.5 shows the worst possible precedence; all parts have to be assembled at a fixed point in the sequence and there is no choice. Looking at the precedence shown in Fig. 3.3, there is considerable choice. An excellent feature can be seen in column 2. Here, not only are there many options but they are also in an early column; the implication here being that choice at an early point of assembly is more important than choice at a later stage.

A very poor feature can be seen in column 4. Here there is no sensible choice and this is almost inevitable since turning the S/A over 'must' be a unique activity.

Figure 3.5 also demonstrates that the term 'stacking' assembly indicating a well-designed product is misleading. What is meant by this is that products which assemble from vertically above with no re-orientation of the S/A are good but that multiple side-by-side assembly is really needed to give a good precedence diagram.

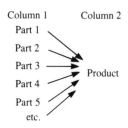

Figure 3.4 Precedence diagram for a product with 'perfect' precedence

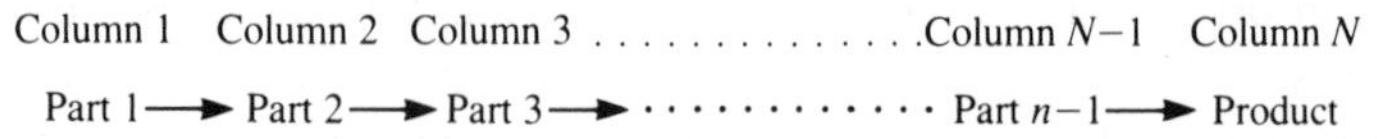

Column 1   Column 2   Column 3 . . . . .Column $N-1$ Column N

Part 1 ⟶ Part 2 ⟶ Part 3 ⟶ · · · · · · · · · · · · Part $n-1$ ⟶ Product

Figure 3.5 Precedence diagram for a product with the worst possible precedence

Sequencing priorities

Because there are conflicting requirements for sequencing it is sensible to indicate the first priorities.

Chapter 2 introduced the concept of failure and considered the fact that significant failures in the assembly process are due only to parts quality, and that there is inevitably a parts quality less than perfect which gives lowest cost assembly. It was also suggested that the way to accommodate failure is to ensure that both failures and the consequence of failures are minimized.

If product design can influence precedences which in turn influence the effect of failures, then it is necessary to investigate what can happen when a failure occurs.

The first requirement for dealing with failures is to be able to detect the failure. The method of failure detection depends on the assembly process; these will be discussed in Chapter 4. It suffices at this stage to say that failure detection is both reliable and inexpensive.

When a failure occurs there are two basic categorizations of possibility: either the partial assembly is scrapped or some form of 'error recovery' can be effected. Since recovery is conditioned by the type of failure, the assembly methods and the assembly process, these will be discussed later. Here, where independent factors are being considered, only scrapped partial assemblies will be discussed. Scrapped assemblies also fall into two categories, those that are scrapped because the product is spoiled, e.g. electric light bulb manufacture, and those that are scrapped because the cost of recovery is greater than the 'value' of the assembly at the point of scrapping. In either case, if scrap is to be minimized then scrap needs to have minimum value; this involves a combination of the number of items scrapped and their individual values. The term 'value' is used because it is in common parlance. The author would maintain that in manufacturing nothing adds value, it only incurs cost. Value is only added when somebody 'buys' the item, i.e. if scrap is to be minimized it needs to have incurred the least cost.

The effect of precedence on cost of scrap is best illustrated by use of a hypothetical example which, for simplicity, is assumed to have perfect precedence and to compare, for this, the best and worst sequences of assembly from the viewpoint of cost of scrap.

Consider the simple example shown in Table 3.1. If all the possible combinations of assembling the parts are considered, the cost of scrap for the various solutions is as shown in Table 3.2.

Table 3.1 Hypothetical example, having perfect precedence

Part No.	Incurred cost* (p)	Defect levels (%)
1	1	1
2	2	1
3	1	2
4	2	2

*For simplicity, the cost assumes that the cost associated with manual assembly is 2p per part and with dedicated assembly 0.2p per part.

Table 3.2 Cost of scrap for hypothetical example

	Assembly cost(p)	Minimum cost of scrap/product	Maximum cost of scrap/product
Manual assembly	8.0	0.44	0.61
Dedicated assembly	0.8	0.18	0.27

The inference from Table 3.2 is that, for any form of assembly, having more control over the assembly sequence will always result in lower costs of scrap but that the potential savings for manual assembly are less because of the production volumes. For dedicated assembly, when the production volumes are high even though the potential savings per product are less, the total savings are significant. For 10 million products per year on an assembly cost of £80 000, for this idealized situation the savings would be £9000.

Another feature that can be related to precedence is that associated with commonality of sub-assemblies for variants of a common product. Perhaps the best example of the influence of this is to be seen in the current practice of customizing products to suit particular valued customers. There are many examples, for all kinds of products, for which the brand name is that of the retailer rather than the supplier. Often the 'branding' is superficial and is only evident in the packaging, but there are increasing numbers of examples of the product itself being identified. In these circumstances, stocks are maintained to satisfy customer demand and in many cases the complete product needs to be stocked because the customized item is assembled at an early stage in the assembly process. It is much more sensible to customize the product at as late a stage as possible so that partial assemblies, common to many customers, can be built and stocked to be finished on receipt of order.

A similar philosophy can be used for variants of a product. It is a common practice for a manufacturer to market a range of products with the same functionality but different prices to suit the market requirement. Again, using common sub-assemblies is effective even when it often means that the less expensive versions of the product are fitted with more upmarket sub-assemblies.

In a wider context, good product design can help to ensure that the most expensive items of a product are assembled last. One of the skills in manufacturing is to incur costs as late as possible and to add value as early as possible. Ideally this would result in costs only being incurred on the same day as payments are received. If assembly is organized such that the costs incurred for expensive items happen as late in the manufacturing cycle as possible, this, in effect, reduces the cost burden carried by the product. An extension of the same argument applies to low-cost items that are required late in the manufacturing cycle. Operating just-in-time strategies for these can easily be counter-productive; the penalty for running out of these is often very high compared with their value.

3.4 Standardization

Perhaps the single most effective activity that can occur in design for assembly which is independent of assembly methods and processes is standardization—not just within a particular product but across the whole manufacturing activity within a company.

Clearly, standardization of product-specific elements is not practicable but there is a wide range of stock items which are duplicated in products. The biggest category of these includes separate fasteners; screws, nuts, bolts, washers, rivets, circlips, spire nuts, etc. Another large group is that associated with sealing typified by 'O' rings, and another significant group is bearings. Too often, designers attempt to achieve minimum functionality within the specification and this can result in two levels of adverse costs. At the first level, the item is still a proprietary one and the actual *direct* cost of the item is hardly affected, but the assembly costs often are. At the second level, the item becomes a 'special' for which the increases in direct cost are significant as well as the increased assembly cost.

Increasing numbers of companies are now identifying and minimizing the number of stock items that they use; if a design calls for items not 'approved', appropriate authority is ended for these items. This approach could be adopted as a matter of course by most manufacturers.

Benefits of standardization

The benefits of standardization are many. Factors independent of assembly include:

- smaller stocks

- less administration of stock

- less storage for stock

- lower cost stock because of increased quantities of specific items

Factors that are assembly-related include:

- less assembly tooling

- less assembly technology

- less development of assembly equipment

- higher confidence in assembly equipment

There are many design features which are independent of assembly methods and processes and which collectively, as well as influencing the cost of assembly, have an effect on manufacturing costs, overheads and product reliability. The most important of these is the adoption of a minimum parts strategy, but benefits can be obtained from appropriate assembly sequencing, enhanced by good precedencing and standardization, both of which are very much conditioned by product design.

CHAPTER 4

Product Design Factors Dependent on Assembly Process

4.1 Introduction

Assembly processes have been broadly introduced in Chapter 2 and it is now necessary to look at these in more detail. Assembly processes fall into three broad categories, manual, dedicated and flexible. Manual assembly can be either serial or parallel, dedicated assembly is always parallel, and flexible assembly can be either serial or parallel. In manual and flexible line assembly, several assembly tasks are carried out at each workstation whereas in dedicated assembly only one task is carried out at each station.

It will be assumed that the overriding factor is assembly cost although, as discussed in Chapter 2, there are many other benefits arising from good design and appropriate methods and processes. If assembly costs are to be reduced, there are only two basic ways in which this can be done. Either more production is needed from the same facility or the same output is needed from a reduced facility, i.e. the overall factor determining assembly costs is the production rate to facility ratio. The higher this is, the less is the cost of assembly.

Facility in this context means the people required, the equipment required and the overheads associated with both people and equipment. In broad terms and within fairly close limits, the production rate for particular products is usually known and this tends to determine the method of assembly, either serial or parallel. Thus for dedicated assembly, when only one product is of concern, improvements in production rate, while welcome, do not have major impact on costs and it is a reduction in facility that gives most of the benefits. For manual and flexible assembly, however, where the facility can always be fully utilized simply by assembling more and different products, increases in production rate as well as reduced cost of the facility are both equally important.

4.2 Factors influencing the production rate to facility ratio

There are only two basic factors that influence the ratio once a method and a process have been identified. These are:

1 the amount of assembly effort required,

2 the effectiveness of assembly effort.

In broad terms, the amount of assembly effort required is independent of methods and processes, and responds only to the number of operations required. This has been discussed at length in the previous chapter.

The effectiveness of the effort responds to far more factors and these will now be discussed in the order in which assembly takes place.

4.3 Parts presentation

It has been said that assembly is a simple activity in that all that is involved is picking something up and putting it down somewhere else. While this somewhat understates the problem it is essentially true *provided* what is being picked up conforms to the restrictions imposed by the picking mechanism and, to a lesser extent, where and how it is being put is also conditioned by the limitations of the mechanism.

It has also been said, with good justification, that the key to successful assembly is efficient, economical parts presentation, i.e. when parts are at the right place, in the right condition, at the right rate, for the right price, most problems of assembly have been resolved. What is right, of course, depends on the method of assembly; how to make it right depends very much on product design.

4.4 Manual assembly

For manual assembly, the operator has significant abilities and what is right for an operator covers a broad range of possibilities. An operator can pick from different locations, can pick parts in different orientations and can manipulate parts very effectively into insertion orientations, i.e. operators can pick most parts from bulk and get them into the right condition relatively easily. The exceptions to this are obvious:

- very small parts

- very large/heavy parts

- difficult to handle parts (sharp, slippery, fragile, etc.)

- parts which are difficult to pick singly (nesting and tangling parts)

- parts with almost identical orientations

In this context, ease or difficulty relates to the time taken for the operation and the exceptions listed above take more time. In a similar way, parts being picked from bulk, having none of the above characteristics but many orientations, take more time to manipulate than parts with few orientations, and, as a consequence, giving parts as much symmetry as possible makes sense. There is sometimes confusion between what is possible and what is required, and it is perhaps unfortunate that people are good at assembly. Ignoring all the points made above would not make parts presentation impossible but it could render it very inefficient. This point will be addressed more fully when discussing flexible assembly.

4.5 Dedicated assembly

For dedicated assembly, the requirements for parts presentation are more stringent. Prior to transfer, a part has to be selected and because the transferring device (Fig. 4.1) is incapable of significant manipulation (changing a variety of orientations into a common one), it is necessary to present parts in a single orientation. Further, because the transfer mechanism operates on a fixed trajectory, it can only pick parts from a single location.

Supplying parts at a single location in a single orientation can be achieved in a variety of ways, but with limited exceptions the normal and most economic method of parts presentation is by the use of an automatic small parts feeder.

There are many examples of this, such as the centre-board hopper (Fig. 4.2), the rotary disc feeder (Fig. 4.3), and the external gate (rivet) hopper (Fig. 4.4), but by far the most common is the vibratory bowl feeder (Fig. 4.5). All small parts feeders have three common features: a hopper for storing bulk parts in random orientation, a separator which takes parts from bulk and an orienter which ultimately presents parts at a convenient point in a single orientation. For the latter, the methodology is essentially

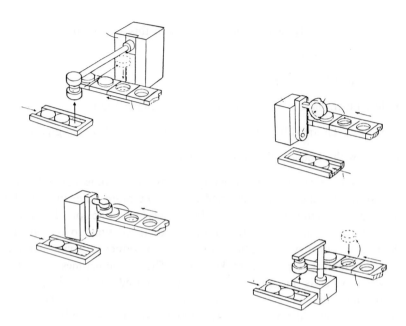

Figure 4.1 Standard transferring (pick-and-place) devices

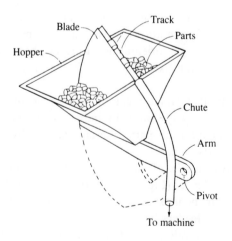

Figure 4.2 A centre-board hopper feeder

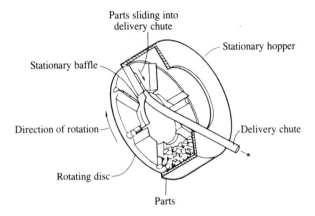

Figure 4.3 A rotary disc feeder

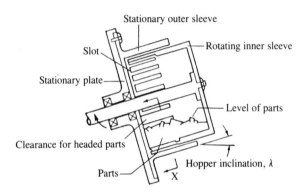

Figure 4.4 An external gate (rivet) hopper feeder

common to all feeders in that parts in a single orientation are selected by rejection of parts in all unwanted orientations. Because the vibratory bowl feeder is the most common small parts feeder, the impact of product design on this will be discussed.

With reference to Fig. 4.5, parts placed in bulk in the feeder are taken from bulk by the vibratory action which conveys them up the inclined vertical spiral track. Before entering the orienting system, mechanical filters (Fig. 4.6a) restrict the parts to single depth, single width before they enter the section with orienting devices proper. The orienting devices then progressively remove (or occasionally convert) unwanted orientations until a single orientation is left; a typical set of devices for a typical part is shown in Fig. 4.6a. For a dedicated assembly system, the rate of

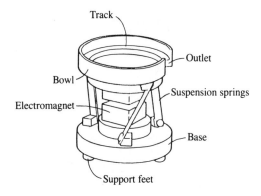

Figure 4.5 A vibratory bowl feeder

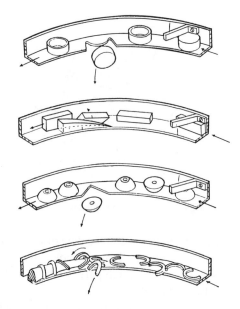

Figure 4.6(a) Examples of orienting systems

requirement of parts is high, say 15–30 per minute. The penalty for lower rates is potentially a loss of efficiency for the whole system since, for all parallel systems, the cycle time is that of the slowest station.

It is useful to examine each element of vibratory feeding to establish good design features.

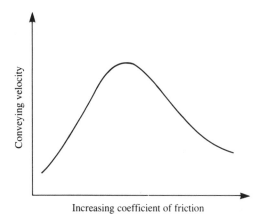

Increasing coefficient of friction

Figure 4.6(b) Relationship between conveying velocity and coefficient of friction for a vibratory bowl feeder

Transportation

The first requirement is that parts will climb the track of the vibratory bowl feeder. This is not normally too difficult to achieve but there are several part properties that are not desirable:

- Low friction between part and track. Figure 4.6b shows the effect on conveying velocity of changing the coefficient of friction between the part and the track; it can be seen that low values of friction result in low conveying velocities.

- Very thin parts. Because the conveying of parts up the inclined track is, at least, partially the result of the assistance (push) received from those travelling around the horizontal base, parts which are 'thin' do not push effectively.

- Abrasive parts. These obviously wear away the track material and this will usually result in changes to critical dimensions.

- Sticky parts. Reference to Fig. 4.6b indicates that high values for coefficient of friction also results in reduced conveying velocity. For sticky parts there are also the problems associated with them sticking to each other.

Separation: single file, single depth

Parts that interact positively with each other such that either force or precise manipulation is required to separate them represent a disadvantage. Because the 'part' as the feeder sees it is larger than a single part, the separation system rejects it. At best, if the parts separate on rejection, this results in a reduction in feed rate; at worst, if the parts do not separate, the feed rate progressively reduces to zero. Parts which interact in the ways outlined above are considered to either 'nest' or 'tangle'.

Orientation

Before looking at the effect of part design on orientation it is useful to discuss the characteristics of typical orienting systems.

- They are invariably fabricated; orienting systems are not normally machined but added to or inserted in bowl feeder tracks, and their geometry and alignment is suspect. This usually means that small feature detection is difficult, which can mean it is difficult to deal with small parts or small features on large parts. Similarly, parts that cannot maintain their geometry (flexible) are difficult to orient. Because of the curvature of the orienting track, prismatic parts 'chord' (Fig. 4.7); this also makes small feature detection for these parts difficult.

- Increasing the conveying speed of parts and hence increasing the rate of parts entering the orienting system does not necessarily increase the output of oriented parts. The loss of output depends primarily on the size of the feature (Fig. 4.8).

- However pronounced a feature might be, it is of no use if it cannot be seen in the direction of conveying.

Additionally, there are other more obvious part characteristics which affect the output of oriented parts:

- Increasing the size of a part—this reduces the output; as mentioned above, for a particular part there is a maximum useful conveying velocity which implies that larger parts are conveyed at a lower rate.

Figure 4.7 Part 'chording' in a vibratory bowl feeder

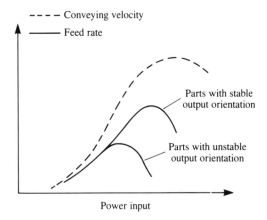

Figure 4.8 Typical relationship between orienting efficiency and conveying velocity for a vibratory bowl feeder

- Reducing the number of orientations—on the basis that all but one orientation will be rejected; less orientations usually implies more output.

- Building in symmetry—two identical orientations usually result in twice the output.

- Minimizing the number of very stable attitudes that the part can take on the track, e.g. significant discs $1 < < d$ or significant rods $d < < 1$ are far preferable to parts with $1 = d$ (Fig. 4.9). This is because the distribution of attitudes depends on the natural stability; the more stable one attitude is compared with the others, the more will be in that attitude at the input to the orienting system.

- Accepting parts in the orientation that give the highest output— changing one orientation into another is neither technically difficult nor expensive.

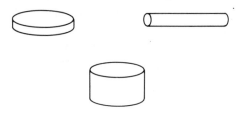

Figure 4.9 Shape significance in minimizing orientations

To summarize, parts should not:

- have low friction between part and track
- be very thin
- be abrasive
- nest or tangle
- be very small
- be very long
- have small orienting features
- be flexible
- have nearly identical orientations

Parts should:

- have as few stable attitudes as possible
- have few orientations
- be accepted in the most common orientation

4.6 Flexible assembly

For flexible assembly, the requirements for parts are virtually identical to those for dedicated assembly with two provisions:

1 The required rate is lower; even for flexible assembly lines there will be multi-tasking per station and the required rate will be at most only 25 per cent of that for dedicated assembly; for single-station flexible assembly the rates will be even lower.

2 The part should be the right way up for insertion. Special devices to change part orientation at the exit to the feeder are costly because they are dedicated to a particular part in a particular product, and they take time to install when the equipment is required to assemble small batches of different products. Alternatively, picking a part, putting it down in a way that changes its orientation, and re-gripping it takes time.

4.7 Gripping

While the quality and scope of gripping might be expressed as the ability to hold a part consistently in the same way with the ability to hold a wide variety of different sizes and shapes of part, the former is not particularly appropriate since gripping, in itself, is not the objective. A better definition of the quality of gripping is the ability to hold a part in a way that allows the part to be inserted; with the proviso, of course, that insertion is possible. Thus, for example, with the former definition of grip quality, people would be poor grippers; with the latter, they can be accurately defined as excellent.

In manual assembly, therefore, with the reservations indicated in parts handling, manual assembly does not have gripping problems primarily because of the ability of people to perform insertion operations despite poorly defined spatial relationships between the mating parts.

In dedicated assembly, a gripper has only to deal with a single part and the scope of the grip, therefore, is not a problem. The quality of the grip, however, is very important since successful insertion operations are primarily the result of positioning. In this section and the sections on flexible assembly, the arguments are based on the use of mechanical grippers, although some of the points are pertinent to other grippers as well. The spatial relationship between a part and its mating part can depend on where the part is, the quality of the grip, the repeatability of the transfer mechanism, and the position of the mating part. In dedicated assembly, the various positions, can, by careful alignment, be well controlled and in this situation it is the quality of the grip which determines the quality of the insertion. It could be argued, therefore, that this is the key to successful insertion. Unfortunately, a good-quality grip is not often achieved in practice for reasons that will now be outlined.

Perhaps the single most common problem in gripping is that there is open tolerancing between the gripped dimension and the insertion dimension. This almost certainly arises because it is not significant for manual assembly yet it is crucial in automated assembly. Good design, therefore, would ensure this tolerance can, at least, be specified.

The second most common problem is that the grip itself is not kinematically sound. The best grip must be a three-point grip whose lines of action are equi-spaced and act through a common point. For nominally axi-symmetric parts this is easy to achieve yet many commercial grippers use, for example, two vees. For prismatic parts, there are more problems. It is easy to say that a four-point contact with one contact a vee (Fig. 4.10) will give the required location, but it is not

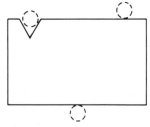

Figure 4.10 A four-point contact grip for a prismatic part

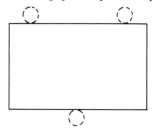

Figure 4.11 A three-point contact grip for a prismatic part

always possible to achieve this. A more common possibility is a three-point grip (Fig. 4.11) where positional errors perpendicular to the direction of grip are possible. These points lead to a specification for the constraints on the part before grip. For an axi-symmetric part, constraining the part is not necessary because the part will be fully constrained by the grip. The part only needs to be located such that it is within the grip envelope with the gripper open; an obvious advantage of this is that the alignment between the feeder and the transfer mechanism is not now critical. For a prismatic three-point grip, the part needs to be constrained before grip in the way that it will not be constrained after grip; this is easy to achieve and should result in a good-quality grip.

The third important point about gripping is the gripping mechanism. Most commercial grippers use two moving jaws for which there must be some doubt as to the position of the centre of grip; more importantly, this can change with time as the gripper mechanisms start to wear. A good-quality grip needs one fixed jaw.

For flexible assembly, there is another dimension to the problem, that of scope. Even on a flexible line it is likely that at least three operations will be carried out using the same manipulator and, hence, three different grips will be needed; for single-station assembly, the number of operations can be many more. Current methods of addressing this problem can be classified as follows.

- *Use a 'universal' gripper*. This is impractical since, to date, attempts to develop a gripper of this type have been unsuccessful. Most efforts have also tried to combine this with manipulative skills presumably because this would be a requirement for selecting parts from random orientations (bin-picking).

- *Use a turret of grippers.* This is a common solution for line assembly with limited tasks; for single-station assembly, in its own right, it has only limited success but can be combined with other strategies. There are also potential access problems with the turret.

- *Use gripper changing.* This is the obvious solution—change the gripper to suit the requirement; the obvious penalty is non-productive time while grippers are being changed. In a typical single-station situation, 30 per cent of total time is due to gripper changing.

- *Carry out multiple assembly.* Since the percentage of time required for gripper changing is clearly based on the average number of parts assembled between changes, if multiple assembly is carried out, the number of gripper changes is fixed while the number of parts assembled increases proportionally to the number of products simultaneously assembled. Most products of the type pertinent to this text are small, and space is not a problem. The penalty faced is that of extra fixture costs.

- *Use special multi-purpose grippers.* This is a common solution where a gripper is designed to be suitable for two or three operations. It is a realistic solution but the disadvantage is that the gripper will inevitably only be suitable for the tasks within the product for which it was designed. How can the product designer help? Clearly the points made for gripping for dedicated assembly are still valid. Additionally, two areas can be given consideration:
  —Design parts such that the grip site is common to all parts. This is clearly not totally practical but it is possible, with sympathetic design, to reduce considerably the number of grippers required for a given number of tasks. The optimum number depends on the gripper strategy adopted, which is part of the specification of the equipment. This is an important point and will be referred to again in Chapter 5 when assembly methods are discussed.
  —Select a sequence of assembly that minimizes gripper changing. This returns to the discussion on precedence and reinforces the view that good design should result in many alternative assembly sequences. It is also yet another different constraint on sequencing which will be discussed later.

4.8 Transferring

The time taken to transfer parts from the part location to the partial assembly is unaffected by product design except in the sense that fewer parts mean fewer transfers. It is, however, a function of assembly methods and will be briefly discussed in the next chapter.

4.9 Part insertion

It has already been said that for successful insertion there is a prerequisite that there has been a good-quality grip. This alone does not guarantee successful insertion and there are clearly design factors that influence insertion. Even though the design factors tend to be very similar for different assembly processes, it is perhaps useful to discuss the differences between manual and automatic insertion.

In manual insertion, the basic insertion action is completely different to that for automatic insertion. In manual insertion the part being inserted is deliberately misaligned so that contact is established between the mating parts. A combination of touch and sight then interact with movement to carry out the operation. By far the most important sense is touch. Three examples illustrate this:

1 Even in blind situations, once a meaningful contact has been made the insertion operation is easy.

2 People are not good at close tolerance work particularly when it is difficult to establish a meaningful contact (Fig. 4.12).

3 Attempts by people to achieve a relatively open tolerance insertion without the mating parts touching are usually unsuccessful.

In automated assembly no touch should be necessary if there is good alignment. The usual problem which invalidates this argument is that,

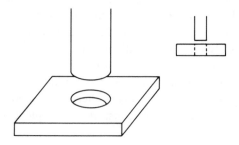

Figure 4.12 A difficult insertion operation for manual assembly

because of a combination of poor-quality gripping, low transfer positional repeatability, variable partial assembly location and unsympathetic tolerancing between the mating parts, these are not correctly aligned. The usual practice is to accommodate the problem using 'touch' type control such as active and passive compliance. It is contended here that often good design can minimize the problem and reduce the requirement for sophisticated solutions.

Unfortunately, in flexible assembly, suggested solutions have usually attempted to mimic how people perform assembly which completely disregards the fact that people do it the way they do because they have no alternative.

Regardless of these arguments, if there is going to be any 'over-design' it is probably best applied to insertion and, in general terms, there are common design rules for all assembly processes:

- Insert from vertically above; gravity is inexpensive.

- Choose as open tolerances as possible.

- Assist alignment by the generous use of chamfers, tapers, etc.

- If a further part needs to be located relative to the current part, ensure that the *final* location of the current part is accurate.

- Design so that the part can be released as soon as possible once insertion has started.

- Do not have more than one insertion site (Fig. 4.13).

- Ensure that the probability of successful insertion is high before releasing the part.

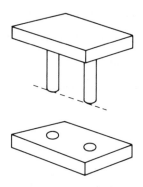

Figure 4.13 Simultaneous insertion in two different assembly locations

- Try to avoid simultaneous insertion operations, e.g. nut and bolt with neither constrained.

- Ensure that there is good access for the gripper.

Additionally, for manual assembly, being able to see the operation is advisable (to establish initial meaningful contact) and, for dedicated assembly, insertion requiring compound motion should be avoided because of cost.

The above has considered the quality of insertion. Additionally, consideration needs to be given to the type of insertion since some operations are inherently more time-consuming or more expensive than others. In order of effectiveness these are:

- insertion from above under gravity,

- insertion from above requiring gripping until the final location is reached,

- insertion requiring compound motion, e.g. translation and relocation,

- insertion requiring excessive motion (the rotation for threaded fasteners),

- insertion from other than above (particularly below).

4.10 Failures

The previous sections have described what is required to successfully carry out assembly operations and the basic assumption has been that mating parts are within tolerances that will allow this. Unfortunately, as has been mentioned earlier, it is not economic to use perfect parts and there will be some failure. These resolve as:

- an unacceptable grasp—no part, wrong part, poor grip,

- inability to start insertion,

- incomplete insertion—usually because of resistance,

- unacceptable insertion—wrong parts, either part added or part present.

When failures occur, it is implicit that some form of recovery is required. In the following sections, the arguments are about what can realistically be done as opposed to what should be done.

4.11 Error recovery

Manual assembly

In manual assembly, the most significant factors are the extensive array of sensors available in the forms of vision, touch and sometimes hearing, and the ability of the assembler to make sensible judgements, based on sensor information, very quickly. With these capabilities, absence of parts, incorrect parts, with gross manufacturing defects, and foreign matter are dealt with either prior to or at the point of handling, and the loss in efficiency is negligible.

For all these circumstances, the action required is to reject the part and no judgement is required other than that associated with identifying the fault. For parts with tolerance defects, an attempt will be made to assemble the part and it is at this stage that judgement becomes important. The possibilities that exist are:

1 The part being inserted cannot reach its final location because:
   (a) the part being inserted is incorrect,
   (b) the part into which insertion is taking place is incorrect.

2 The part reaches its final location but does not give a satisfactory assembly.

The latter is normally detected by either sight or touch and can be quickly dealt with, but the former will usually require further action. This usually takes the form of trying another part on the basis that the probability of selecting two faulty parts consecutively is very small and that if assembly still cannot be achieved, it is the part that is already in the assembly that is faulty. If the part being inserted is faulty then the time lost is that of rejecting the part, but if this is not the case, then a decision has to be made as to what to do and when to do it. This will be discussed in the next chapter.

The actions described above are obvious when the appropriate capabilities are present but, clearly, to achieve the same type of capabilities in automatic equipment is at best expensive and at worst impossible.

Dedicated assembly

In dedicated assembly, comprehensive system-tending is justifiable because of the very high rates of production, and this can be used to effect. For example, automatic recovery from parts-feeding errors is not usually necessary because, provided adequate warning is given that parts are not available for picking, the fault can be rectified before the problem affects the assembly machine, i.e. there is a buffer between the parts feeder and the assembly machine that allows assembly to continue in the event of a parts-feeder failure.

Fault-sensing on dedicated assembly machines is not usually capable of detecting all potential faults, but is effective enough to be considered satisfactory. The sensors used are invariably very simple mechanical or optical binary devices, where a change of state of the device, coupled with compliance in the insertion direction of the workhead, can be used to determine faults. Once a fault has been detected, the action that follows depends on the assembly method and will be discussed in the next chapter. On dedicated assembly machines there is never any attempt at on-line error recovery.

It can be observed from the above that manual and automatic assembly tackle the problem of minimizing the effect of faulty parts in completely different ways; this is based not on sensing the error but on the ability to recover from the error. It is this that needs to be addressed in flexible assembly, i.e. what capability does the robot manipulator have for error recovery and how can this be best used to provide an acceptable solution to the problem?

Flexible assembly

In the history of flexible assembly there has been an unjustifiable compulsion to give this equipment 'human' capabilities, presumably because, in motion terms, the equipment has human attributes. Although this is a very interesting exercise, it assumes that the aim is not to assemble a product in a cost-effective way, but to assemble a product in the same way as people in a cost-effective way; this leads to considerable problems. It would be more relevant but not completely valid to use the argument that robot assembly systems should 'mimic' automatic assembly machines because these, when appropriate, are far more effective than people at assembly.

Parts presentation

Parts are normally presented to a robot gripper by automatic parts feeders or magazines and the location of parts is known and acceptable before being picked. Absence of parts at a feeder outlet could be dealt with as for dedicated assembly by prior warning.

Part acquisition (gripper)

Part acquisition using mechanical grippers gives some capability for inspection, and hence for error recovery. The possibilities are for malformed part, out-of-tolerance part and incorrectly gripped part. With the exception of out-of-tolerance parts, the majority of these can be detected by having a means of sensing the amount of grip, preferably in the end-effector rather than the gripper. The appropriate action in this case is obvious; the part is discarded and another part acquired. Increasing the effectiveness of sensing at this stage is not considered to be cost-effective.

Part insertion

By using dedicated assembly methodologies, most insertion errors can be detected at low cost but, for many, on-line recovery would be either expensive or impossible. If it is assumed that insertion is from vertically above, the two main possibilities are that the part is finally positioned under the action of gravity or the part is finally positioned by some positive action.

Final position by gravity represents the most common insertion operation in which, to save time, the part is released as soon as it has positive horizontal location. If the part does not reach its final location, re-acquiring it may be possible, but would not be easy because its vertical position is unknown. If the part cannot be re-acquired, removal of the partial assembly is mandatory. When the part can be removed it is not known if the part being removed is defective or its mating part in the assembly is defective. This can only be tested by rejecting the part and acquiring and inserting another part. If insertion is now successful, the assembly process can continue; if it is again unsuccessful, it can reasonably be assumed that the mating part is defective and the actions necessary to recover from this situation on-line are considerable. Parts need to be disassembled and stored for further use; extra tooling may be necessary; judgements need to be made about the probability of success; permanently fastened parts present a virtually insurmountable problem; and the programming effort required would be extensive and expensive. The alternative for all insertion faults is to remove the partial assembly for some level of manual rectification or scrap, and this would almost certainly be the most appropriate action.

Final positioning of parts by positive action includes any form of interference fit and mechanical fastening operations and, for these, re-acquiring the part is not relevant because the error is detected before the part is released. The initial problem, therefore, is the ability to remove only the gripped part, since in some circumstances this would not be possible without unjustifiable expensive tooling or fixturing. The appropriate actions, therefore, can be considered to be:

1 If the single part cannot be consistently disassembled do not attempt on-line recovery.

2 If the single part can be removed, reject the part, acquire another part and use the methodology described for gravity-located parts.

The above has suggested what can be done. What should be done depends on the assembly method and will be discussed in the next chapter.

Surprisingly, considering the completely different methods of acquiring, manipulating and inserting parts, there are many common features of good design for the various assembly processes. It has been argued with some justification that, with few exceptions, parts design for assembly processes is process independent and that, for manual assembly, slight over-design for automated assembly is appropriate.

CHAPTER 5

Product Design Factors Dependent on Assembly Method

5.1 Introduction

Assembly methods can be broadly categorized as:

1 Single-station assembly—all the assembly functions are carried out in a 'single' location.

2 Line assembly—only one task is carried out in a given location and the product is assembled by moving the partially completed product from location to location.

3 Hybrid assembly—multiple tasks are performed at a number of locations.

Using these definitions, manual bench assembly and single-arm flexible assembly are single-station assembly, dedicated assembly is line assembly and all other methods of assembly are hybrid. For example, one person assembling a product where a series of partial products are moved in turn to the operator for several assembly tasks to be performed, is hybrid assembly. Similarly, flexible line assembly where the partial product passes through the system having several assembly tasks performed at each location is also hybrid assembly.

In Chapter 4, failures were discussed for various assembly processes and the conclusions can be summarized as follows:

- For manual assembly, error recovery is easy if the failed part is the one being inserted, but if the failure is caused by the part already present then error recovery is possible but potentially expensive.

- For dedicated assembly, error detection is basically easy but error recovery is ostensibly impossible.

- For flexible assembly, error detection is usually possible and some error recovery can also be achieved. However, many errors cannot be recovered.

It has also been suggested previously that errors are primarily caused by attempting to assemble faulty parts, that these parts cannot be economically eliminated and that dealing with these successfully relies on minimizing their effect on overall performance.

In Chapter 2, the situation in which an error results in the partial assembly being scrapped was discussed; it was concluded that good product design resulting in 'good' assembly precedences can minimize the effect (cost) of scrap. In this chapter, the effect of the assembly method on performance will be discussed and particular reference will be made to recovery from assembly errors.

5.2 Single-station assembly

In single-station assembly, if it is the part being added that is faulty, then, if recovery is possible, it is clear that the faulty part should be removed and replaced. If either recovery is not possible because the part already present is damaged or the part already present is faulty then, again, what to do is obvious. The product needs to be disassembled to the point at which the faulty (damaged) part can be removed, after which, assembly can recommence. In manual assembly, the mechanism for achieving this is straightforward; both the disassembly work and the assembly work is carried out by the operator. In flexible assembly, as discussed in Chapter 4, disassembly has to be carried out by an operator and assembly from the point at which the defective part has been removed can either be carried out manually or by using the flexible assembly equipment; for the latter there are some system supervisory problems but none that cannot be overcome.

Knowing the mechanisms of recovery, what can be done to minimize the effort required to carry out the recovery? If the problem is associated with the part being added and it is immediately recoverable, then better product design, from the standpoint of the assembly method, cannot improve the situation. If, however, this is not the case, then the amount of remedial action can be influenced by the order in which the product is assembled. Obviously, since all the inefficiency is caused by the amount of disassembly work required, assembly should be arranged such that this is minimized. This can be done by choosing an order of assembly in which the two mating parts are either assembled as soon after each other as possible, or retrieving the offending part with minimum disruption (if it is not 'buried' under other parts).

By implication, single-station flexible assembly is carried out in a single-assembly location where the product is built up and then removed on completion. With all automated assembly it is probable that not all the operations can be carried out by the equipment. This could be due to a difficult parts-feeding operation, a difficult insertion task, a difficult grasping operation or a particularly poor-quality part that would benefit from manual inspection. This implies that the partial assembly may have to be removed and then reintroduced into the system, which means that either the partial assembly has to be removed from the fixture and reinstated on the fixture, or there need to be extra fixtures. The former results in wasted time and the latter results in extra cost. It is possible that sympathetic product design can resolve this by:

- ensuring that manual operations are at the beginning or the end of the assembly work and hence not relevant to the machine,

- carrying out the manual operations off-line by making small sub-assemblies which can be introduced to the machine as parts.

Clearly, the latter is the more realistic possibility and the success of this will depend on how 'open' the precedences are and how soon parts are secured after they have been placed. Both of these aspects of product design have arisen in other contexts earlier.

5.3 Line assembly

Before the effect of product design on line assembly can be discussed, it is pertinent to comment on the various forms of line assembly and to compare the performance of these. As stated previously, line assembly is invariably restricted to dedicated assembly machines which are broadly classified as indexing (synchronous) or free transfer (asynchronous); these are shown diagrammatically in Figs 5.1 and 5.2. Indexing machines, as the name implies, operate by moving *all* the assembly fixtures simultaneously and it is implicit that an event which stops one workstation operating stops them all. Conversely, free-transfer machines link independent workstations by buffers in which partial assemblies, on fixtures, may be stored. With this system of operation, a stoppage at any individual workstation does not immediately or necessarily stop any of the other workstations from functioning.

In the event of an error, if it is accepted that the error cannot be

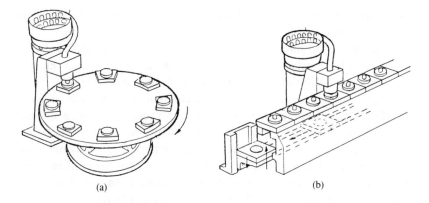

Figure 5.1 Indexing (synchronous) assembly machines (a) rotary, (b) in-line

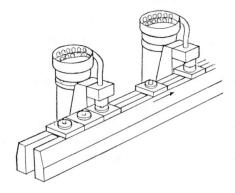

Figure 5.2 A free-transfer (asynchronous) assembly machine

corrected automatically, three possibilities exist. The first is to stop the machine and attempt to recover from the error manually—in this case, if the recovery is successful the machine can continue. If the recovery is not successful (a current part is not causing the problem), the product would invariably be removed and downstream stations would sense that there was no partial assembly on the fixture and hence not carry out an assembly activity.

The second possibility is to acknowledge the error, allow the machine to continue, do no further work on the partial assembly and carry out manual remedial work at the end of the assembly line.

The third possibility essentially changes what is done in the remedial work phase and allows for reintroduction into the line of a reworked sub-assembly. This possibility is really only appropriate to free-transfer machines but in the authors' experience it is never done. It can be concluded, therefore, that a combination of little demand, extra capital equipment cost and low potential savings are the reasons for this.

Invariably, free-transfer machines stop when an error is detected in the hope that the part being added has caused the error, the rectification time will be short and the disruption to the system minimal. Indexing machines sometimes stop but more often continue and do no further work on the partial assembly. The former are usually referred to as 'stopping' indexing machines, the latter 'memory pin' machines. Figure 5.3 shows the relationship between production rate and parts quality for typical 'stopping' indexing, 'memory pin' and free-transfer assembly machines. It is assumed that the machines all work on a cycle time of three seconds and that, for the machines that stop, it takes 30 seconds to rectify the fault and restart the machine. It can be seen from the figure that the highest production rate of acceptable assemblies is obtained by the 'memory pin' machine and the lowest by the 'stopping' indexing machine. For the 'memory pin' machine, this production rate is not achieved without a cost; this is that of the rectification work required for the unacceptable assemblies that this type of equipment produces.

For 'stopping' indexing and free-transfer machines, all 'current' part faults are rectified when they occur and only 'previous' part faults reach the end of the machine. For these, two factors need to be considered:

1 As for single-station assembly, the position in the assembly of the current part relative to the faulty 'prior' part has to be considered: the further they are apart, the more disassembly work is required.

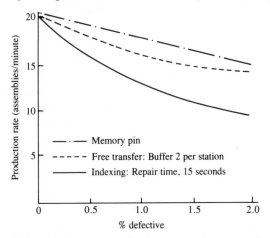

Figure 5.3 Relationship between production rate and parts quality for the various types of assembly machines (for a 10-station machine with a cycle time of three seconds)

2 The nearer the end of the assembly the faulty 'previous' part, the less manual assembly is required to finish the product. Simplistically, this means that parts which are known to cause 'prior' part faults should be added as late as possible. This, of course, assumes that completing the assembly is the most cost-effective thing to do. Under circumstances where this type of fault unavoidably occurs early in the assembly sequence, it is more effective to completely disassemble the product and return the good parts to their parts feeders. In general terms, therefore, 'prior' part faults should preferably be as near the end of the sequence as possible but if this is not possible, they should be near the beginning of assembly.

For 'memory pin' machines, the strategy for dealing with 'prior' part faults is exactly the same as above but additionally, when a 'current' part fault occurs, the partial assembly passes to the end of the line with no further parts added. Since there are invariably many more of these than there are 'prior' part faults, they represent a much more serious problem. Unfortunately, there is no better way of accommodating these than the methods used for 'prior' part faults, i.e. they need to occur either early in the assembly sequence or, preferably, as late as possible so that manual disassembly or assembly respectively is minimized. Inherent in this is product design for 'open' precedences in which individual activities are as independent of each other as is possible.

The above has considered how the effort, and cost, of dealing with unacceptable assemblies can be reduced by good product design. The second aspect of efficient assembly is the production rate of acceptable assemblies.

Figure 5.3 shows the relationship between production rate and parts quality for typical conditions and where, for simplicity, all part matings in the assembly are of equal quality. Obviously in a real situation this is not the case; some part-to-part combinations are of good quality while others are less good. Given a free hand, where should parts be placed in the assembly sequence to minimize the effect on the production rate of unacceptable assemblies? This will depend on the assembly method.

For a 'stopping' indexing machine, the production rate of acceptable assemblies is only affected by the number of 'current' and 'prior' part faults and the time it takes to rectify a 'current' part fault. All of these points are independent of the assembly sequence and hence nothing can be done in the current context to improve production rate.

For a 'memory pin' machine, the production rate of acceptable assemblies is also independent of assembly sequence since all faults are certain to occur and they always result in unacceptable assemblies.

◯ Stations

▢ Buffer spaces

Figure 5.4 Schematic of open-loop free-transfer machine

For a free-transfer machine, the situation is different because of the independence of the individual workstations. Most free-transfer machines are open loop as shown in Fig. 5.4, i.e. regardless of the size of the buffers between stations, the first station never stops because it has no empty assembly fixture available and the last station never stops because it has no buffer location in which to place an empty assembly fixture. As a consequence of this, the machine is biased towards the middle in that the stoppage of a station near the centre of the machine is more significant than one towards the ends, i.e. if there is a choice of where to place poor-quality part combinations, they should be at the ends rather than the middle of the machine. Fortunately this is consistent with the strategy for limiting the problems associated with producing unacceptable assemblies, and there is no conflict.

The third aspect of efficient line assembly is the accommodation of manual operations. As mentioned previously, most assembly work is such that some operations need to be carried out manually. For automatic assembly lines, this does not represent a significant problem; all partial assemblies are carried on fixtures and the introduction of operators at appropriate positions on the line does not cause any disruption. There are, therefore, no product design implications in this when considering the assembly method.

5.4 Hybrid systems

While, in theory, virtually all assembly systems have elements of at least two assembly methods, genuine hybrid systems are of three basic types which are:

1 Manual assembly lines where several tasks are performed at each station on the line.

2 Single-station manual assembly using an indexing mechanism and multiple fixtures, and with one operator assembling multiples of products.

3 Flexible assembly lines where several tasks are performed at each station on the line.

Manual assembly lines

Perhaps the single most important aspect of manual line assembly is balancing—that is, giving equal effort to all operators on the line. In dedicated lines with one operation per station, it has to be accepted that the system can be no quicker than the slowest operation and, once the operations have been optimized—often by good product design—nothing can be done to increase the theoretical production rate. In manual assembly, however, two things ultimately determine how well the line is balanced; the first is the size of the individual tasks that make up the assembly operations, and the second is the ease or difficulty with which individual tasks can be moved to different positions within the assembly sequence. The former can obviously be influenced by good design, as was suggested in Chapter 3; the latter can be influenced by good design by ensuring that the product has 'open' precedences. This is yet another example of the importance of choice in the sequence in which operations are carried out.

Manual assembly lines are of three basic types: they are either machine-paced indexing, operator-paced indexing or free transfer. A machine-paced line is one where the transfer device operates at predetermined intervals, and it is the task of the operators to ensure that their portion of the work is finished before transfer. Clearly with this method of operation, fluctuations in individual operator performances are bound to occur and to accommodate this, the lines are fitted with buffers, i.e. the operators are to some extent working in isolation and their mean performance determines how well the line balance operates. An operator-paced line, as the name implies, is controlled by the operator such that the transfer device only operates on receipt of a signal from each of the operators, which indicates that they have all finished their tasks; this method does not use buffers. Clearly, if this method of operation is to be successful, an effective incentive scheme is necessary. A third possibility operates essentially like a dedicated free-transfer machine. Operators work with upstream and downstream buffers and sub-assemblies are released when a downstream buffer space is empty. All the systems mentioned are well used, but currently, perhaps the most popular is the latter, primarily because it is relatively easy to reconfigure the line and to increase or decrease the size of the line.

Product design influences the operation of all these systems in the same way as it does for indexing and free-transfer dedicated lines, and as outlined earlier in the chapter. The important extra element in manual line assembly, however, is the capacity for rework which does not exist on dedicated lines, i.e. on dedicated lines anything that passes an assembly location proceeds to the end of the line before anything further can be done to it. For a manual, operator-paced indexing line which has no buffers, the same is essentially true but for lines with buffers, there is scope for rework of 'prior' part faults at individual workstations and also 'prior' part faults by transferring back along the line a reworked sub-assembly to an upstream buffer. The latter, while both useful and possible, is better than error recovery in dedicated assembly in that the reworked sub-assembly is reintroduced into the line; it is not to be recommended, however. Rework at an individual station has two obvious advantages: it is local, and, by implication, the amount of disassembly needed is reduced.

Good product design for manual line assembly should reflect this by ensuring, wherever possible, that 'prior' part faults are contained within a single workstation, i.e. part matings, particularly if they are subject to faulty assembly resulting from the 'prior' part, should be near to each other in the assembly sequence. Unfortunately, this contradicts a design feature discussed previously which indicated that combining parts to give multi-functionality is effective. Since the reduction in operations is basic to good design for assembly, this rule must be followed, but with a rider. The rider states that when a part has multi-functionality, parts and operations that rely on it should be added as soon as possible and particular care should be taken to ensure that those which cause the largest numbers of faulty assemblies be dealt with first.

Single-station manual assembly with indexing fixtures

This form of assembly, which is currently very popular, attempts to exploit the best features of line and bench assembly. An indexing table with a fixture at every location has one assembly station. An operator, at this location, carries out a number of tasks (typically the same amount of work content as an operator line), the machine indexes and the operator repeats the operations; this continues until the first work carrier returns, when a second set of tasks is performed, etc. The parts supply necessary for every assembly cycle is automatically presented in appropriate locations.

Clearly, with this method of assembly the benefits of repetitions of limited assembly operations, together with the benefits of giving the operator meaningful work, are both present.

Because this is effectively a modified version of single-station assembly, the rules for effective assembly and error recovery are the same and are as stated earlier in the chapter.

Flexible assembly lines

There are two basic types of flexible assembly lines:

1 very large lines with up to four tasks per station and, by implication, many tasks, and

2 relatively small lines with rather more tasks per station but relatively few tasks.

Large flexible lines

These essentially operate in a similar way to dedicated assembly lines with two important differences: they can deal very effectively with variants and they produce intermediate annual production volumes. This type of equipment has been exploited very effectively, particularly by Japanese industry, for relatively expensive consumer products. With this type of equipment there are usually, at the most, limited buffers and importantly, the line is essentially not reconfigurable, i.e. product runs are long and the concept of small batch production of significantly different products is not possible.

The general rules of automated line assembly for efficient running are valid. As with all forms of automated assembly, automatic error recovery is not possible nor is it desirable, but, as for single-station flexible assembly, reintroduction of reworked partial assemblies is possible. Essentially, therefore, this form of assembly relates directly to buffered manual lines in that relationships between parts should be as close as possible in the assembly sequence and preferably within the scope of one assembly location.

As was the case for dedicated lines, the requirement for carrying out manual operations between automated activities does not have a serious cost or time implication on the performance of the line. The only small performance implication of this results from the probable imbalance between the manual and automated activities. This leads to a rider which states that manual tasks should, where possible, be grouped together into portions of work which have the same assembly time as the cycle time of the equipment.

Small flexible lines

These are essentially linked single-station flexible assembly machines with the following advantages:

1 They produce higher production rates.

2 They overcome the problem of presenting large numbers of parts to a single assembly location.

With this type of equipment, as for single-station flexible assembly, the aim is to perform small-batch assembly of significantly different products. Unfortunately, most examples of this form of assembly and single-station assembly do not meet the above criteria. They offer only assembly variants for a long production period, hence can use dedicated feeding equipment relatively economically. Perhaps the only example of a genuine small-batch producer is a prototype piece of equipment which has been developed jointly by a European consortium of academic and industrial partners.

This is shown in Fig. 5.5 where it can be seen that the machine consists of three main zones: a parts supply zone, a main assembly zone and a power operations zone. The assembly zones are provided with all their needs by a common materials handling system for large parts, fixtures, tools, etc. and up to 24 linear vibratory feeders for small, easy-to-handle parts. The intention is that for typical sub-assemblies, the equipment will be capable of assembling different variants with no reconfiguration time, and for significantly different products, the reconfiguration time would

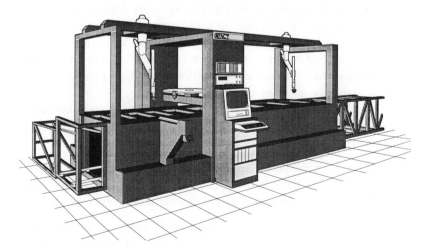

Figure. 5.5 Flexible assembly system for small batch production

be less than one machine cycle (typically five minutes). Furthermore, and importantly, the reconfiguration costs are minimized by using standard pallets, low-cost feeder tracks (see Chapter 6), and a range of standard grippers and tools. The latter is achieved by assembling in multiples of the sub-assembly on a common fixture. This ensures that gripper and tool change time is minimized and hence simple reusable tools and grippers can be used.

To help reduce the assembly time per operation, assembly in the main manipulator zone is carried out at two different locations. This at least partially overcomes the problem of the distance between grasping and insertion associated with assembling many parts in a fixed location and also, as a bonus, helps to keep the main manipulator active at all times. The power operations zone, as the name implies, is for carrying out power operations. The manipulator requirements for this type of operation, typified by high torque nut or screw running, need a manipulator with different characteristics than that of the main manipulator which is essentially a versatile, high-speed-pick-and-place unit. Because there will be an unavoidable imbalance between the two assembly zones and there will generally be fewer, longer operations in the power zone, only one fixture location is used in this area.

This approach to the flexible assembly of small batches of significantly different sub-assemblies has a marked advantage over other approaches; it specifies a piece of equipment which would be suitable for many different tasks rather than designing equipment which is only appropriate for a limited range of tasks. The product design implications of this are profound. If the equipment, which can take many different forms for this type of assembly, is 'designed to suit', it is very difficult to formulate meaningful design rules. If, however, the equipment is fixed, design rules can be established and product designers can use this knowledge to effect. This exercise then becomes similar to design for other forms of assembly and to design for primary manufacture.

Most of the design rules pertinent to this method of assembly are rules that are merely applications of those already discussed and can be summarized as follows:

- Deal with unacceptable assemblies in the same way as dedicated machines; since there is no mechanism for automatic rework, do no further work on the partial assembly once a fault has been detected.

- Ensure as far as possible that a 'prior' part which causes an unacceptable operation is as near to the operation as possible if the unacceptable assembly is to be reworked and reintroduced into the machine.

- If unacceptable assemblies are to be scrapped, put poor-quality part combinations as near the beginning of assembly as possible.

- If unacceptable assemblies are to be reworked and completed manually, put poor-quality part combinations as near the end of assembly as possible.

- Try to organize the assembly sequence such that necessary manual operations are carried out either before or after introduction to the machine. If, as is likely, this is not possible, form sub-assemblies which include the manual operations and introduce these to the machine as parts.

Rules specific to the equipment are:

- Try to keep the partial assembly in one location as far as is possible; moving between fixture locations too often overworks the materials handling system and will eventually result in this controlling the cycle time of the equipment. Segregating pick-and-place operations and power operations is clearly one manifestation of this.

- Try to avoid using palletized parts immediately after each other; this also tends to overwork the materials handling system. This has been designed to be low-cost working on the principle of a shared resource being able to meet the overall materials handling requirements; it is much less effective and hence inefficient when confronted by a localized 'peak' demand.

There is no problem associated with design for gripping but clearly a need for less grippers has cost implications and should not be completely ignored.

This completes this three-part discussion of product design, having covered that which is independent of the assembly method and the assembly process, that dependent on the assembly process, and that dependent on the assembly method. Chapter 6 looks at various commercial and prototype design-for-assembly systems and considers the relative merits of alternative systems; it is followed by three chapters featuring case studies for three different DFA methods. Finally, Chapter 10 examines the possibilities for design for assembly at the conceptual design stage and indicates, through examples, some lines of approach that might be adopted.

PART III

Design-for-Assembly Methods

CHAPTER 6

Design-for-Assembly Methodologies

6.1 Introduction

Time-based competition, i.e. the ability to quickly develop, produce and distribute new products and processes, has become increasingly important as a strategic means of competition, as well as progressive development and diversification to meet customer needs. Rapid changes in markets and technology tend to bring shorter product life cycles.

Developing products quickly is not, however, a goal in itself. Only in those cases where the finished products result in a competitive advantage for the company are reduced lead times useful. Lead time is also strongly correlated with two other very important achievement standards, cost and quality, and these must also be considered. It has been suggested that it is not an issue of reductions in lead time *per se* that are important but reductions in lead time for a given level of resources and design quality, which result in more competitive manufacturing.

Product development includes all the activities that take place from an interpretation of market needs and technical possibilities to finished production designs, i.e. drawings, specifications, tools, fixtures and production programs. Also included in product development are prototype production and test activities.

Product development is part of the technical innovation process. The innovation process includes all activities from the generation of ideas to first commercial production.

6.2 Approaches to product development

A number of organizational solutions are recommended to improve communication and cooperation between the participants in technical innovation processes. For product development especially, a number of such solutions are known. For example, integrated product development, in which the separate tasks overlap, and multi-functional project groups have been studied. Two fundamental approaches to product development, i.e. sequential and integrated, are discussed here.

Sequential approach to product development

The traditional way of developing products is similar to a relay race in which the development, like the baton, is passed on in a stepwise fashion. Such a process is often divided into a number of distinct phases executed in a predetermined order, i.e. concept development, feasibility testing, product design, prototype building, process development, pilot production and final production (Fig. 6.1).

When a team of functional experts have finished their work the 'baton', i.e. the responsibility, is handed to those in charge of the next phase in the development chain. This sequential procedure, in which project responsibility is handed over successively, is often called 'over-the-wall'.

A sequential approach to development work leads to highly differentiated and functionally specialized tasks. This approach has some advantages. The development becomes easy to manage and control since the objective of each phase is predetermined and the process is scrutinized after each phase is finished. This procedure reduces uncertainty to a minimum before each phase is begun.

The sequential process has been criticized as being ill-equipped for advancing such development objectives as speed and flexibility.

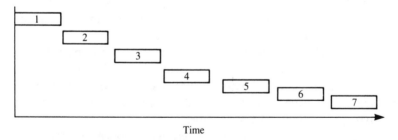

Time

Figure 6.1 Sequential approach to product development

Integrated product development

Alternative development processes have evolved. They have names such as synonymous, simultaneous, parallel and overlapping. The basic idea behind all these concepts is the same. By letting the various functions overlap to some extent, the total development cycle is reduced while at the same time flexibility increases (Fig. 6.2).

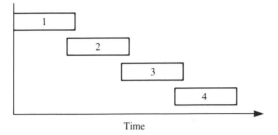

Time

Figure 6.2 Overlapping phases of product development

This is achieved through the use of cross-functional (multi-disciplinary) project groups which at the beginning of the project include representatives from most functions, i.e. marketing, engineering, manufacturing, quality and field service. The goal is to open the channels of communication between the participants as early as possible. By effective communication, the downstream tasks can begin based on preliminary data, which means that time can be gained in later stages.

Simultaneous, or concurrent, engineering is an additional aspect of integrated work which deserves mention. It is an engineering-oriented procedure with focus on the interactions between design and production. The concept refers to the simultaneous activities by which the production process is adjusted at the same time as the product is developed.

An example of a successful simultaneous product development is the new sports-car engine GM Chevrolet-Pontiac Canada Group, Detroit, MI. By applying the precepts of simultaneous engineering, the total product development time was reduced from seven to four years. Figure 6.3 illustrates a comparison between traditional and simultaneous development processes for the LT-5 engine.

6.3 Approaches to design for assembly

The objective of design for manufacture (DFM) is the integration of product design and process planning into one common activity. DFM embraces some underlying principles which help maintain communication between all elements of the manufacturing system and permit flexibility to adopt and modify the design during each stage of the product's realization.

Design for assembly (DFA) as a central element of DFM has one important characteristic; it addresses product structure simplification, since the total number of parts in a product is a key indicator of product assembly quality.

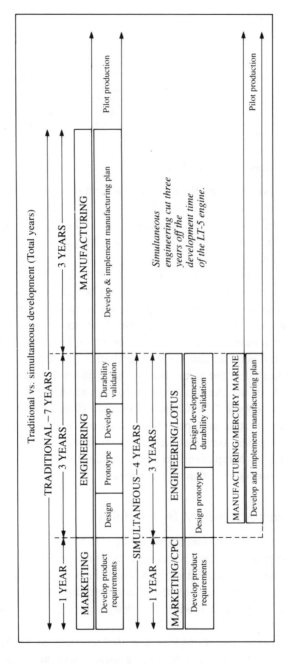

Figure 6.3 Sequential versus simultaneous approach to product development (*Source*: *Machine Design*, March 1990)

A number of different DFA methodologies have been developed. What properties should these DFA methods have that are of interest to designers? They need to be:

1 Complete—the DFA method should have two complementary parts:
   (a) objectivity, i.e. procedures for evaluating assemblability, and
   (b) creativity, i.e. procedures for improving assemblability.
   The majority of DFA methods only have a strong objective part. For a designer it is important to know how to change or influence factors since knowing that things are wrong does not naturally lead to things that are right.

2 Systematic—this characteristic indicates that the methodology involves a systematic, step-by-step procedure, which helps to ensure that all relevant issues are considered, i.e. the organization of objective and creative parts of DFA methods.

3 Measurable—one of the central problems of DFA is how to measure assemblability objectively, accurately and completely. Current DFA methods use different assemblability measures. Cost information is the most objective measure of product assembly quality. The goal of assemblability evaluation is to find the optimal combination of influence factors. The accuracy of cost estimation is an indicator of the quality of the DFA method. Alternatively, it can be argued that a knowledge of whether things are improving is sufficient and that this will logically lead to the lowest cost solution. Unfortunately, this line of reasoning has two disadvantages: firstly, it does not allow the necessary interaction with other design criteria such as design for manufacturability, since without cost data the necessary compromises cannot be made. Secondly, and critically, since the cost is not known, there is no indication of whether the design will lead to a product which is successful commercially.

4 User-friendly—the user-friendliness of any DFA methodology is critically important as regards implementation cost and designer effort. A very fine balance is necessary between the ease of use and the quality of the design exercises. If a particular method is easy to use but results in virtually no benefits then it is not a good system. Conversely, however 'good' a method is, if it is very difficult to learn and/or use then it will not be used.

A major barrier to DFA is usually time. To reduce training and evaluation time, the DFA method must be presented in an effective form. Design and manufacturing engineers are typically operating to very tight schedules and are, therefore, reluctant to spend time learning a DFA method. DFA methods are presented as handbooks, monographs, evaluation procedures with spreadsheets or as computer-aided systems.

Bearing in mind the above-defined requirement, current DFA methodologies can be classified as being one of four basic types:

1 Those that have design principles and design rules.

2 Those that employ quantitative evaluation procedures.

3 Those that use a knowledge-based approach.

4 Those that use computer-aided methods.

6.4 DFA systems using design principles and design rules

Conversation with any manufacturing engineer responsible for assembly indicates that there are countless examples of good and bad designs from the viewpoint of assembly. Collectively these designs offer a wealth of data which can be used to convert assembly knowledge into design principles, design rules and guidelines.

This human-oriented knowledge helps to narrow the range of possibilities for good principles and rules, but the skill in creating a good DFA method is in identifying the relative importance of the various pieces of data, resolving conflicts between the data and organizing the resulting information so that it is usable.

Design rules are empirical truths verified by extensive design practice. Many design rules have been created, some of which are very general while others have more detail. Andreasen (1983, 1985) has published work on:

• product assortment

• product structure

• parts

This work is useful as an introduction to design for assembly.

A very comprehensive catalogue of rules have been created by Weissmantel (1989), who categorizes rules into the various areas of product concept, product structure, joining techniques, etc. The rules are illustrated by many graphical examples.

There are many other DFA methods that catalogue design rules; for reference to these, see the bibliography at the back of this book. Most of the methods have more or less the same rules interpreted in a variety of different ways.

An example DFA method using design principles

Suh (1982, 1988a, 1988b) has developed design principles that are generally valid laws of good design. His belief is that there are fundamental design principles or axioms, the use of which to guide and evaluate design decisions leads to good design. This approach consists of two basic axioms with derived theorems and corollaries:

1 *Independence* Maintain the independence of functional requirements. This applies strictly to feasible designs, i.e. those that satisfy the functional requirements.

2 *Information* Minimize information content; this depends on the first axiom, i.e. information content is to be minimized subject to the fulfilment of functional requirements.

By definition, axioms must be applicable to the full range of manufacturing decisions. From the axioms many consequences have been developed. Corollaries may pertain to the entire manufacturing system or may concern only a part of manufacturing system. Examples of corollaries are:

1 Decouple or separate parts or aspects of a solution if functional requirements are coupled or become coupled in the design of products and processes.

2 Integrate functional requirements into a single physical part or solution if they can be independently satisfied in the proposed solution.

3 Minimize the number of functional requirements and constraints.

4 Use standardized or interchangeable parts whenever possible.

5 Make use of symmetry to reduce the information content.

6 Conserve materials and energy.

7 Reduce the number of parts.

Measurement of information content is based on probability theory. The definition of the measure of information is:

$$I = ln \frac{L_1}{[\Delta L]}$$

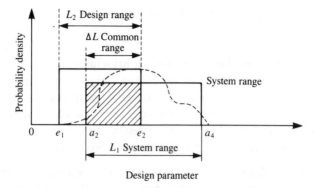

Design parameter

Figure 6.4 The information axiom and its implication

where L_1 is the system range and ΔL is the overlap between the design range and the system range (Fig. 6.4).

The axiomatic approach has been applied to many problems. Similarities to the axiomatic approach can be found in the Taguchi method of design for quality.

6.5 DFA systems employing quantitative evaluation procedures

Quantitative evaluation procedures allow the designers to rate the assemblability of their product designs quantitatively. These quantitative DFA methodologies are systematic and this ensures that the DFA rules are being correctly applied and the influence factors are being correctly evaluated and improved. They stimulate a creativity and usually reward the designers with an improved quantitative rating if they do well. With frequent usage, the need for further use is reduced.

DFA methods favoured in industry are almost exclusively based on evaluation procedures. The designer has to determine the assembly process operation by operation. Each assembly operation is subject to a rating that measures how easily the process can be carried out by operators or assembly systems. For the product as a whole, a quantitative measure is calculated which combines the individual ratings by a formula. The designer can improve the assemblability measure by redesigning those parts that caused bad ratings.

There are several well-known quantitative evaluation methods. Perhaps the best known DFA methods are the Boothroyd–Dewhurst method, the Hitachi method and the IPA Stuttgart method. The Hitachi method and the Boothroyd–Dewhurst method will be discussed in more detail in Chapters 7 and 9 respectively.

IPA Stuttgart method

The Fraunhofer Research Institute, Stuttgart has developed the IPA Stuttgart method. The method is based on evaluation of assemblability of product and improvement of product during the whole design process. Assembly-oriented design processes are defined and technical requirements are systematically implemented into the design process at various stages (Fig. 6.5).

The assembly-oriented design process shows that, among the large number of applicable aids, the most important are design rules and the evaluation of assemblability. During the various stages of the design process, different rules can influence the product design as regards assembly. After completing each stage of the design process, an evaluation of its suitability for assembly is determined. An assembly-oriented review will provide information on the weak points of the product that need to be improved.

Design rules

Design rules in the IPA method are summarized in a catalogue, which is divided into four categories:

1 Measures for the product structure,

2 Measures for sub-assemblies,

3 Measures for individual parts,

4 Measures for joining techniques.

Evaluating suitability of products for assembly

Evaluating the suitability for assembly of products in relation to the different stages of the design process requires that the most important factors of influence at each stage be evaluated. The IPA method allows an evaluation of those factors that are important in conceptual and preliminary design stages. For this, check-lists are used. If a more detailed design is available a more detailed evaluation procedure can be used. The procedures are based on value analysis; the relationship between the functional content of a part and the required assembly expenditure is established. The efficiency of an assembly procedure is improved when parts with high functional content require a considerable assembly effort.

The main steps of the IPA method are:

1 The setting up of a functional structure

AIDS FOR ASSEMBLY-ORIENTED PRODUCT DESIGN / STEPS IN THE DESIGN PROCESS	CREATIVE						CORRECTIVE				
	Catalogue with guidelines	Catalogue with examples	Catalogue of relative cost	List of component use	Competition analysis	Assembly-priority graph	Calculation aids	Value analysis	Analysis of use value	Check list	Evaluation procedure
1. Planning Stage											
1.1 Recognition of requirements	●				●		●	●	●		
1.2 Definition of tasks					●						
1.3 Create a requirements list										●	
1.4 Release for rough design											
2. Rough Design											
2.1 Analysis of functions	●				●						
2.2 Create the functional structure	●	●								●	
2.3 Create variations of the functional structure	●	●								●	
2.4 Determine solutions for each function	●	●							●		
2.5 Select principles											
2.6 Prepare solution variants	●										
2.7 Work out different concepts	●			●		●					
2.8 Evaluate and select the different concepts	●					● ●	●	●	●	●	● ●
3. Drafting Stage											
3.1 Drafting the main functional units	●	●	●	●							
3.2 Drafting the remaining functional units	●	●	●	●							
3.3 Select suitable parts of the draft	●	●					●	●	●	●	●
3.4 Detailed design of the main and auxiliary functional units	●			●							●
3.5 Check and improve drafts											
3.6 Trace faults and problems							●				
3.7 Analysis of cost recovery											●
3.8 Draft completion	●			●			●		●	●	
3.9 Decision on draft						●	●		●		
4. Final Design											
4.1 Detailing	●			●							
4.2 Working out specifications	●			●							
4.3 Analysis of production data											
4.4 Release for production							●				●

Figure 6.5 Assembly-oriented design process

2 The weighting of functions and the determination of the functional content,

3 The setting up of a sub-assembly structure,

4 The allocation of functional contents to the different parts,

5 The determination of the assembly sequence,

6 The determination of assembly expenditure,

7 The determination of values measuring the suitability for assembly,

8 The identification of technical problems concerning assembly expenditure.

This procedure makes it possible to evaluate different product designs at different stages of detail. Figure 6.6 shows points considered during the design process. The influence factors are penalized according to the difficulty of assembly. Figure 6.7 shows a portion of an evaluation chart for size and weight.

The IPA Stuttgart method is systematic. The catalogue of design rules can be used for redesign.

A new version of the IPA Stuttgart method uses assembly costs per part as a quantitative measure of the assemblability. Total assembly costs include manipulation, presentation costs and insertion costs.

Many other DFA methods have been developed—most of them based on the Boothroyd–Dewhurst method. A good overview of DFA methodologies is given in the bibliography.

6.6 DFA methods employing a knowledge-based approach

The rapid development in artificial intelligence has brought knowledge-based systems to the engineering community, providing an environment for storage and inferencing knowledge. Knowledge-based technology allows the user to do some things now for which there is not a good algorithmic base.

Knowledge-based systems are defined as those that provide new information-processing capabilities such as inference, knowledge-based management, search mechanisms, etc., combined with conventional computer capabilities. A knowledge-based system comprises a knowledge base, inference, communications and knowledge acquisition (Fig. 6.8).

Knowledge-based processing for assembly has the following features:

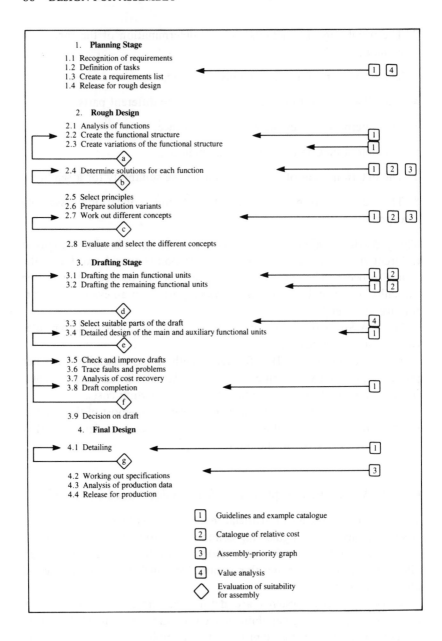

Figure 6.6 Evaluation procedures at various stages of design

Number	Factor	Values	Penalties Manual assembly	Penalties Automated assembly
1	Part size	< 150 mm 150–300 mm > 300 mm	1.0 0.5 2.0	1.0 0.5 3.0
2	Part weight	G < 3 kg 3–8 kg > 8 kg	1.0 2.0 4.0	1.0 2.0 3.0

Figure 6.7 Influence factor penalties

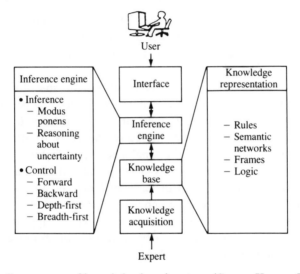

Figure 6.8 Components of knowledge-based systems (*Source*: Krause, Lehmann)

1 Expressions can be stored to allow knowledge to accumulate and be used for problem-solving later.

2 Things can be described that are not known precisely in advance, i.e. it is possible to describe a hypothesis.

3 Advice on the consequences of design decisions on assembly costs can be obtained and suggestions for redesign given.

4 Explanations of questions and deductions can be given.

5 The declarative language of knowledge-based systems is more 'comprehensive' than a procedural language because natural languages are mostly declarative.

Knowledge representation

The knowledge base contains knowledge of an application domain. The central problem of knowledge processing is acquiring a good knowledge representation. Rules, semantic networks, frames and logical expressions are available. Rules describe 'if-dependent' actions. These are widely used and offer a basis for successful knowledge-based systems. An example of a rule is the rule for representation of orientation efficiency knowledge:

> **Rule**: If rotation symmetry *and* concealed end-to-end asymmetry *and* mass is not at geometric centre *and* part envelope is a long cylinder *then* possibly *method* orient using gap tooling: manufacturing time 15 hours.

Semantic networks and frames can be used for structured representation of objects. A semantic net is a relational connection of objects. Objects can be handled as nodes, their relations as edges. Nodes and edges are provided with names.

A frame is a description of an object and contains so-called slots for all information which may be associated with the object. Slots may be store values or pointers to other frames, groups of rules and procedures. Frames have the ability to send and receive messages. Another feature of frames is inheritance. This is the capability to give features of frames to succeeding frames. These features can be modified for inheritance. To represent the designer's world of terms and objects, an object-oriented concept is being developed for a knowledge-based working environment. The aim is to create a working environment which supports individual design processing for different design tasks. An example of frame representation of orienting knowledge is:

Class:	Product
Frame:	Part
Instance of:	Product
Slot: attributes:	Name, form, size, weight, moment of inertia, orientation efficiency
Attribute:	Orientation efficiency
Type:	Real
Value:	Action evaluate results (orientation)
Result:	Orientation
Rule:	
Slot	if: part ... form [rotational symmetry]
	part ... form [concealed end-to-end asymmetry]
	part ... weight [is at geometric centre]
	part ... envelope [long-cylinder]
	then: set (part, orientation efficiency, 0.5)

The value of the attribute of orientation efficiency is attained by searching the rules.

Logic-oriented representations describe knowledge by logical expressions. The most common forms are logic by expressions and predicates. This approach has been applied to the representation of design axioms in an expert system. The Lucas DFA method using inference mechanisms is described in some detail in Chapter 8.

6.7 Computer-aided DFA methods

DFA by conventional or knowledge processing involves consultation sessions in which users have to reply to many questions on part geometry, size, insertion processes, etc. Currently, to reduce user input, assemblability evaluation processes are being developed by which DFA systems are integrated with CAD. The key role in this integrated process is the representation of technical objects and procedures for extraction and processing of assemblability attributes from 3D CAD models. This will now be described.

The product data model

In assembly, a product data model is acquired from:

- a part data model

- the assembly model

Each of these models can be complemented by operation data, for example the number of pieces, the surface-finish data, etc.

6.8 The part model

It is not possible to assign explicitly all the data required by DFA to the part model without an extra workload for the designer. It is necessary to obtain this information implicitly, for example, by automatic calculation of moments of inertia, size, etc. (the geometric model of a part does not have assembly and manufacturing content). A key element for extracting assemblability data from 3D CAD part models and incorporating them into a DFA system is feature processing.

6.9 Feature processing

A feature is a partial form or product characteristic which is considered as an entity and which has a semantic meaning in design, process planning, manufacture, assembly, cost estimation or other engineering disciplines (FEM (finite element methods), etc.). Features are logic connections of form features and semantics. Form features are groups of contours, surfaces, volumes, etc. Manufacturing features are defined for machinability and assemblability; design features express functions. Figure 6.9 gives a definition and examples of features; Fig. 6.10 shows ways of modelling features.

Design-by-features approach to DFA

In the design-by-features approach, features are incorporated in the part from the beginning. Generic feature definitions are placed in a library from which features are instanced by specifying dimensions and location parameters and various attributes (Fig. 6.11).

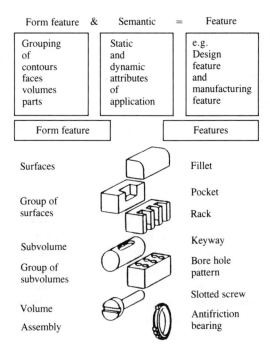

Figure 6.9 Interpretation of features

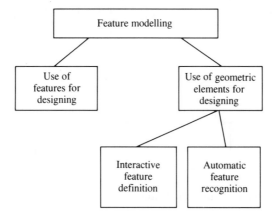

Figure 6.10 Feature modelling

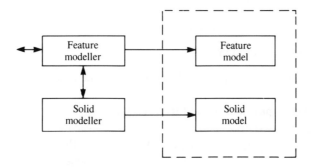

Figure 6.11 Design by features

Interactive feature definition approach to DFA

This approach is based on creating, first, a geometric part model. Features are then defined by human users picking entities on an image of the part. This approach is illustrated in Fig. 6.12.

Automatic feature recognition approach to DFA

A geometric model is created first, and then a computer program processes the database to discover and extract features automatically. This approach is illustrated in Fig. 6.13.

The methods that are used to recognize features can be categorized as follows:

1 Syntactic pattern recognition

2 State transition diagrams

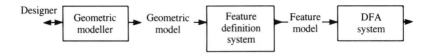

Figure 6.12 Schematic interactive feature definition

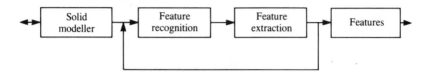

Figure 6.13 Schematic automatic feature recognition

3 Decomposition approach

4 Expert systems approach

5 Constructive solid approach

6 Graph-based approach

For illustration, the graph-based approach will be described.

Graph-based approach to feature recognition

In the graph-based approach, the boundary representation of the part is converted into a graph $G = (N, A, T)$, where N is the set of nodes, A is the set of arcs and T is the set of attributes assigned to the arc. Joshi (1987) has developed a graph (AAG) to represent features. Here, each face of the part is represented as a node, and each edge on face adjoined is represented as an arc. The attributes take values of either 0 or 1 if the two adjacent faces are concave or convex, respectively. Recognition of features is done by analysis of the attribute adjacency graph. Figure 6.14 shows the AAG for a pocket feature. The assumption made here is that the features are made of depressions or cavities. A production rule is

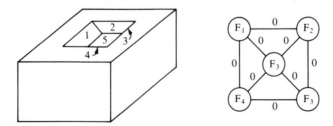

Figure 6.14 Example of pocket recognized by the rule

formulated for each feature type based on the properties of the feature graph. A sample rule for recognition of a pocket for orientability evaluation can be as follows:

Rule: *If* the graph is cyclic *and* it has exactly one node n with number of incident 0 arcs equal to total number of nodes − 1 *and* all other nodes have degree = 3 *and* the number of 0 arcs is greater than the number of 1 arcs (after deleting node n) *then* the corresponding feature is a pocket.

6.10 The part model and assemblability evaluation

The part model can give useful data for the assemblability evaluation such as:

- shape symmetry
- feature symmetry
- dimensions
- weight
- centre of mass
- moment of inertia
- surface-grasping location, type, surface finish, size
  —joining
  —functional
- orthogonality
- colinearity
- parallelism
- features—chamfers, etc.

Many algorithms have been developed for extraction of assemblability data for DFA systems. Prototype algorithms have been developed for extraction of part dimensions and part symmetry for use as an input to a DFA system. A knowledge-based system has been developed for determining grasping properties from a CAD part data model.

Very interesting systems have been developed for automatic parts feeding. From an automatic analysis of part data models with reference to orientability the system generates the optimal type and arrangement of orienting devices for the track of a linear vibratory conveyor. The system is linked to a CAPP (computer-aided production planning) system for the production of the devices. Figure 6.15 shows an example of the CAD model of a part and the appropriate orienting devices.

Feature processing uses the AHPC (automatic handling parts code) system for identification of features important in assembly and generates a group-technology-like five-digit handling code. An algorithm called the feature point attribute symmetry technique (FPAS) reduces a complex part to a set of geometric attributed points in space which represent the features in the body. The example in Fig. 6.16 shows a part which is rotational and has a protrusion on both the side and a face and two steps. The generated code has been designed to be the same as that produced by the Boothroyd–Dewhurst method.

6.11 The assembly model

The model of the assembly represents the relationships between parts in the assembly where the assembly is a collection of bodies that are spatially related.

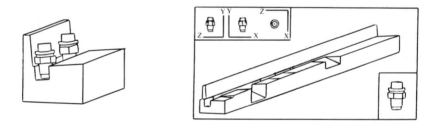

Figure 6.15 CAD/CAM system for the design and manufacture of orienting devices

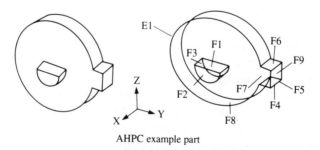

AHPC example part

Figure 6.16 Orientability evaluation by feature processing

Part interaction models can give the following data for an assemblability evaluation:

1 Location and orientation of parts

2 Assembly directions

3 Tolerancing

4 Collision-free assembly/disassembly processes

5 Neighbourhoods

6 Degree of freedom of connections

7 Stabilities

Researchers have developed a group-theoretic approach to representation of assembly. They define the symmetry of shapes and their features. A symmetry of a feature is simply a rotation or translation that maps the feature to itself. This approach has been used to generate an assembly-task specification for robots. Another approach has developed a system for automatic assembly which is capable of dealing with 3D structures and substructures. Representation of assembly is on a CSG (constructive solid geometry) tree and connectivity graphs.

A further approach has resulted in the development of an algorithm for the generation of all the mechanical assembly sequences for a given product. The algorithm takes a description of the product and returns the corresponding AND/OR graphical representation of assembly sequences; assembly sequences generated in this way can be evaluated. The first criterion is to maximize the number of different sequences in which the assembly tasks can be executed. The second criterion is to minimize the total assembly time by simultaneous execution of assembly tasks.

Another knowledge-based DFA system (ADAM) has been developed for use with CAD with the emphasis on the minimizing of part count, the rationalization of standard parts and design for insertion.

A novel development has been that of interactive graphical disassembly as a tool for evaluating the ease of assembly. The design engineer loads the CAD model that specifies the assembly task and, part by part, disassembly is simulated. A part (or sub-assembly) and a disassembly direction is selected. Based on the disassembly process, the system classifies the operation by the selection of a symbol. Every selection is supported by the system with default choices. This computation is based on rules containing generic knowledge about assembly generally and about the current product specifically. By disassembling the product part by part, one assembly sequence is discovered. Other sequences and sub-assemblies are discovered by selecting an alternative part at a particular point in the analysis.

6.12 Assemblability measures

Assemblability tends to mean different things to different people. Currently used measures of assemblability of products are detailed below.

In principle, the following are used:

1 Qualitative measures

2 Quantitative measures

Qualitative assemblability measures

Qualitative assembly information can be obtained through cost structures, design rules and relative costs. These methods are used to find the cost, the influencing factors or the least expensive solution within a group of alternatives.

- Cost structures are a division of costs into several identifiable parts. The parts can be assigned either absolute or relative values. Cost structures are used to determine the types of costs which have the largest influence on costs.

- Design rules are probably the most widely used type of qualitative assemblability measure. They contain the experiences gained from previous design and assembly-process exercises and form the basis of design for assembly. Design rules help the designer to avoid expensive solutions.

- Relative costs are costs for an activity in proportion to the costs for some reference situation. Relative costs can be determined for different types of activity by evaluation or experience. The main advantage of relative costs is that they provide a fast and easy way to find a less expensive alternative. Figure 6.17 shows the results of work that determined relative assembly costs for some screw fasteners.

The advantages of a qualitative assemblability measure are:

1 It is available with different levels of detail.

2 It can avoid the need to estimate absolute costs.

3 It is company independent to a considerable extent and it is hardly subject to change over time.

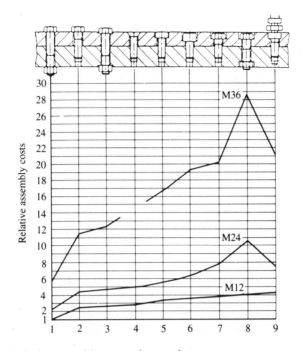

Figure 6.17 Relative assembly costs of screw fasteners

Two disadvantages are:

1 For the designer it is difficult to remember hundreds of rules; information must be accessed at the right time and quickly.

2 Some rules are not well defined; it is not clear when they may be used and the reason for their existence is not obvious.

Quantitative assemblability measures
Quantitative assemblability measures are used to measure the assembly quality of a product in terms of:

1 Global assembly costs

2 Detailed assembly costs

3 Assembly indices

4 Assembly pointers

5 Information

Global assembly costs

Global assembly costs are always presented for particular classes of activity features. Very restricted classes can be formed in which the cost information is valid for activities that are similar to some basic activities. In all cases the information is derived from costs for existing activities in the class.

Statistical techniques such as regression or correlation analysis are used to develop cost-estimating relationships contained in the parametric model. The IPA Stuttgart method, for example, uses global cost functions for the estimation of robot assembly costs. These cost functions were created from cost information for existing assembly robots and regression analysis was used to determine weighting factors.

This type of assemblability measure is applied when it is difficult to perform a detailed cost estimate or when high accuracy is not required.

Detailed assembly costs

Assembly costs are determined based on an estimate of operation times and hourly rates for assembly equipment, operators, etc. The detailed assembly cost estimates require complete product data and a known (or assumed) assembly method. This type of assemblability measure requires a lot of knowledge about assembly processes, assembly process planning structures of assembly systems, etc. The Boothroyd–Dewhurst method uses this type of assemblability measure. The estimation of detailed assembly costs must be calculated using real current data. Changes will inevitably occur. These changes can be financial (changing labour rates, interest rates, etc.) or technical (new faster curing adhesives, new improved equipment, etc.). The method for estimation of detailed assembly costs should be updated regularly to reflect these changes.

Global or detailed assembly cost estimation methods can tell the user how much a design will cost or how much can be saved by specific changes in the design. Target assembly costs can be checked or alternatives can be compared.

Assembly costs are company dependent, i.e. they have to be calculated by the company. Assemblability information which can lead to high potential savings will not often be applicable to other companies.

Assembly indices

Assembly costs are the most objective measure for product assembly quality. Although the ultimate target is cost reduction, assembly cost alone cannot express whether the design quality is good enough, or identify the causes of bad assemblability; a quality index is also required.

An index can indicate the difficulty of an assembly operation, a grasp or an orientation. It can be related to different measures of difficulty, e.g. assembly operation cost or time. The Lucas DFA method uses this type of measure. The Hitachi method uses a penalty score as one assemblability measure. Figure 6.18 shows a fragment of a table for determining the assemblability index for different assembly operations.

Assembly pointers
Quantification of assemblability by assigning points to influencing factors is a frequently used method expressing assemblability by one measure. Each influence factor is evaluated by group specialists who, according to some appropriate methodology, assign points to factors. Total rating for a product is then calculated from the summation of the points. Figure 6.19 shows information for determining points for particular design parameters.

The advantage of this type of assemblability measure is speed and ease of quantification. If the range is wide enough, points can be used to aid understanding of influence factors. The disadvantage of the points method is the difficulty in achieving unambiguous interpretation by different people for the same rule.

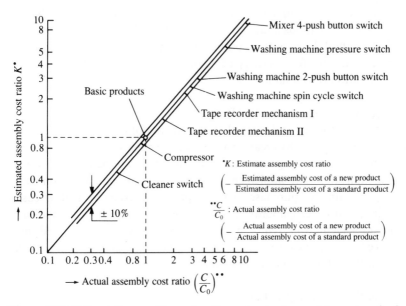

Figure 6.18 Estimated assembly cost ratio and actual assembly cost ratio for small products

		PART POINTS				
NO.	CRITERION	0	1	2	4	8
C1	Weight	0.1 g < G ≤ 2 kg	0.01 g ≤ G ≤ 0.1 g 2 kg < G ≤ 6 kg	G < 0.01 g 6 kg < G		
C2	Size	5 mm < L ≤ 0.5 m	2 mm ≤ L ≤ 5 mm 0.5 m < L ≤ 2 m	L < 2 mm 2 m < L		
C3	Delivery	Oriented		In bulk		Single packed
C4	Separation	Easy in mech. way			Possible in mech. way	Tangle together
C5	Orientation	Easy in mech. way			Possible in mech. way	Imposs. to handle aut.
C6	Flexibility	Inelastic			Poss. to handle aut.	Imposs. to handle aut.

Figure 6.19 Point values for assessing assemblability

Information

The algorithmic approach to DFA uses the amount of information required as a general measure of assemblability. A design is considered better if, on rationalization of functional requirements, less information is needed.

6.13 Guide to DFA comparisons

The above has attempted to give an insight into the principles behind various DFA methods and, inevitably, a potential user will want to know which method best suits a particular purpose. Figure 6.20 shows a comparison table for a variety of methods and the following is a key to interpretation of the table.

1 Implementation cost and effort

A rating of 'better' indicates that the methodology can be effectively implemented simply by creating awareness through seminars and/or brief training, and by management expectation that it will be used. A 'worse' rating indicates that implementation may require extensive company-wide commitment, purchase or development of extensive and/or costly training, and other elaborate preparations. An average rating indicates that implementation requirements are relatively uncertain and may involve varying degrees of software expense, training expense and organizational and procedural changes.

2 Training and/or practice

A rating of 'better' indicates that relatively little training or practice is required for effective use. A 'worse' rating indicates that extensive

Applications	A, B C, D, E	A	A,B,C D,E,F	A,B,C D,E,F	A, B	A,B,C D,E,F	C, E	A,B,C D,E,F	A
Disadvantages	A, B		G	E		H	E, F		C
Other advantages	A, B E, F	A, E H	D	D, E, F	A, E	A, E	F G, H	G, H	A, C E, F
Teaches good practice	●	●	◐	●	●	●	●	◐	●
Quantitative	○	●	●	●	●	◐	●	●	○
Systematic	◐	●	●	●	●	◐	●	◐	○
Stimulates creativity	●	●	◐	○	●	●	◐	◐	●
Rapidly effective	●	●	◐	●	○	○	◐	◐	●
Product planning team approach	●	◐	●	○	●	●	◐	◐	○
Management effort	●	◐	○	●	◐	○	●	●	●
Designer effort	●	○	○	○	○	○	◐	◐	●
Training and/or practice	◐	◐	○	●	●	◐	◐	◐	●
Implementation cost and effort	●	●	◐	●	●	●	◐	◐	●
Criteria / Method	Design axioms	DFA method	Taguchi method	Knowledge-based DFM	Hitachi method	Value analysis	Manuf. proc. des. rules	Computer-aided DFA	DFA, guidelines, rules

● Better   ◐ Average   ○ Worse

Figure 6.20 Comparison table for design-for-assembly methodologies

training and/or extensive user experience is needed; in other words, effective use is directly dependent on effective training. An 'average' rating indicates that a significant commitment to training and/or extensive practice in using the method is required.

3 Designer effort

A rating of 'better' indicates that little or no additional designer time and/or effort is required to make effective use of the methodology. A 'worse' rating indicates that significant additional design time must be allocated for use of the methodology. An 'average' rating indicates that a designer must make a commitment to using the methodology, that perseverance may be required and that some additional design time needs to be allocated.

4 Management effort

A rating of 'better' indicates that little or no management effort or expectation is required. A 'worse' rating indicates that significant management effort and commitment are required and/or that effective use is directly dependent on management expectation and support. An 'average' rating indicates that successful use of the method requires management expectation that the method will be used, coupled with good support in using the method.

5 Product planning/team approach

A rating of 'better' indicates effective product planning and/or the team approach. A 'worse' rating indicates that the methodology can or may require considerable effort to foster good planning and/or the team approach, depending on circumstances.

6 Rapidly effective

A rating of 'better' indicates that the method is likely to be rapidly effective in producing beneficial results. A 'worse' rating indicates that benefits are not likely to be apparent for some considerable time and that the use of the methodology, therefore, requires a long-term view. An 'average' rating indicates that the methodology can or may be rapidly effective depending on circumstances.

7 Simulates creativity

A rating of 'better' indicates that effective use of the methodology tends to require and stimulate design innovation and creativity. A 'worse' rating indicates that the use of the methodology in itself is not likely to influence design creativity. An 'average' rating indicates that design creativity will be stimulated and probably required if the methodology is to be applied successfully.

8 Systematic

A rating of 'better' indicates that the methodology involves a systematic, step-by-step procedure which helps to ensure that all relevant issues are considered. A 'worse' rating indicates that the procedure is not exhaustive and that the degree of 'completeness' is left to the user. An 'average' rating indicates that some aspects of the methodology are systematic and some rely on user interpretation.

9 Quantitative

A rating of 'better' indicates that the method is primarily quantitative in nature and that one or more design ratings are generated. A rating of 'worse' indicates that the method is primarily qualitative. An 'average' rating indicates that the method has both quantitative and qualitative features.

10 Teaches good practice
A rating of 'better' indicates that use of the methodology teaches good design-for-assembly practice and that formal reliance on the method will diminish with use. A 'worse' rating indicates that use of the methodology in itself does not significantly improve the design-for-assembly capability of the user and that the benefits produced depend on continuing formal use of the methodology. An 'average' rating indicates that the method teaches good practice but must still be formally applied if all potential benefits are to be obtained.

Advantages
A key to the advantages listed in the table is as follows:

A Narrow range of possibilities
B Results in inherent 'robustness'
C Ready reference to best practice
D Emphasizes effect of variation
E Helps identify and prioritize corrective action
F Provides both guidance and evaluation
G Can shorten design/tooling cycle
H Can reduce tooling and fixturing costs

Disadvantages
A key to the disadvantages listed in the table is as follows:

A Interpretation not always simple
B Requires 'buy-in' on the part of the user
C Exceptions are not indicated
D Rates only ease of assembly and does not address parts handling or other relevant assembly parameters
E Development requires input from experienced experts familiar with the method's capabilities and needs
F To be used on a regular basis, implementation must be 'designer friendly'
G Must be developed and/or customized for each specific application
H Often requires difficult-to-obtain information

Applications

A key to applications appropriate to each method is as follows (some of the methods catalogued are general design aids and they are not specific to assembly):

A Mechanical and electro-mechanical devices and assemblies
B Electronic devices and systems
C Manufacturing process and other processes
D Software, instrumentation and control, systems integration
E Material transformation process
F Specialized and/or unique manufacturing facilities such as flexible assembly systems

6.14 Conclusion

This review of well-known DFA methodologies has been made with the focus on the principle and measures of assemblability. The three basic approaches to DFA have been identified as:

1 Design principles and rules

2 Quantitative evaluation methods

3 Knowledge-based systems

Each approach has been discussed, and well-known DFA methods for each approach explained. Different assemblability measures have been identified. Advantages and disadvantages of assemblability measures have been presented.

It is generally considered that knowledge processing mixed with conventional processing and CAD technology is the direction for the future.

In the next three chapters, case studies are presented, one for each of the different methods identified above.

CHAPTER 7

The Hitachi Assemblability Evaluation Method

Hitachi Ltd developed their Assemblability Evaluation Method (AEM) as an effective tool to improve design quality for better assemblability. Since then, it has been widely used by the Hitachi Group and by many other national and international companies, and is considered to be one of the more effective methodologies. The work on AEM started in 1976 (coincidentally around the time that the Boothroyd–Dewhurst methods were first being developed) and, over the next few years, the methods were refined in the light of new and different experiences. Later versions of AEM included many additional elements, the most notable of which was the inclusion of evaluation procedures for the assembly cost of individual parts, i.e. it was found that a purely comparative exercise, while very useful, was not sufficient—realistic cost information was also necessary.

7.1 Objectives of AEM

The main objective of AEM is to facilitate design improvements by identifying 'weakness' in the design at the earliest possible stage in the design process, by the use of two indicators:

1 An assemblability evaluation score ratio, E, used to assess design quality by determining the difficulty of operations,

2 An assembly cost ratio, K, used to project elements of assembly cost.

Hitachi regard it important to consider both the cost and the quality of assembly since, for example, a low-cost design might easily not be the lowest cost design and, conversely, an excellent design might be too expensive because of the particular circumstances.

Figure 7.1 shows the information flow for the evaluation of

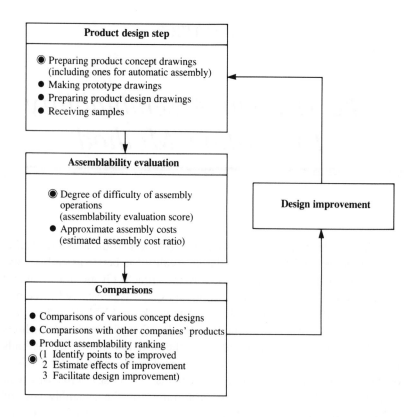

Figure 7.1 Assemblability evaluation and design improvement flow diagram

assemblability and the subsequent design improvement. The first stage of the evaluation procedure requires conceptual drawings but, in common with other methodologies, the more comprehensive the initial information, the more effective is the result; completed product design drawings and prototypes will give better results. This partly negates the effectiveness of design for assembly in that, ideally, design should be 'right' as early as possible in the design process; this aspect of design will be discussed in Chapter 10. As with other DFA methods, the Hitachi method is particularly appropriate for comparative studies and can be used to make reasonable judgements about the quality of in-house or competitors' products.

Design improvement is achieved by reviewing and interpreting the evaluation results and by then carrying out further assemblability evaluation; this iteration continues until the design has been optimized.

7.2 Theory of evaluation

AEM is based on the following procedures:

- Assembly operations are first categorized into approximately 20 elemental assembly tasks and each task is assigned a symbol which clearly indicates the content of the task. These tasks relate specifically to insertion and fastening processes and not to parts handling. It is argued that because at this stage of design the important issue is the spatial relationship between individual parts, assembly sequencing is not an important factor, and the concept of automated assembly is difficult to evaluate; these issues are considered later. It is further argued that for manual assembly, parts handling is a secondary issue which could detract from the more important insertion considerations.

- Each of the elemental tasks is subject to a penalty score which reflects the degree of difficulty of the task. The penalty scores are obtained from analysis of shop-floor data and, importantly, are constantly revised to reflect changes in technology and methods. The penalty scores are then ranked and all are compared to the elemental task with the lowest penalty score. Some examples of elemental operations and penalty scores are shown in Fig. 7.2.

- Factors which influence elemental tasks are extracted as coefficients and the penalty scores are modified accordingly.

- Attaching (contacting) conditions appropriate for each part are expressed using further AEM symbols.

- The sum of the various penalty scores for a part are then modified

Elemental operation		AEM symbol X	Penalty score e_x
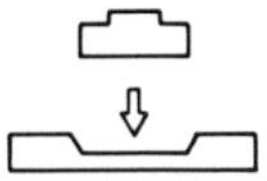	Downward movement	↓	0
	Soldering	S	20

Figure 7.2 Examples of the AEM symbols and penalty scores

by the attaching coefficients and subtracted from the best possible score (100 points) to give the assemblability evaluation score for the part.

- The total assemblability evaluation score for the product is now defined as the sum of the assemblability scores for the individual tasks, divided by the number of tasks. This now may be considered to be a measure of design efficiency where a score of 100 would represent a perfect design. Hitachi consider that an overall score of 80 is acceptable.

The assemblability evaluation score in itself does not provide all the information necessary to determine the advantage to be gained by reducing the number of parts. Indeed, it is possible to improve the assemblability evaluation score by increasing the number of parts, if the parts being added have a higher-than-average assemblability evaluation score. Clearly, this is inconsistent with the objectives and to overcome this, a cost ratio is used. The assembly cost ratio is defined as the assembly cost of the redesigned product to the assembly cost of the original product, and is a function of the number of parts in the product.

7.3 Accuracy of AEM

Estimated assembly cost ratios are continuously compared to actual assembly cost ratios once these have been established in production. If the deviation is small (better than five per cent) then it is considered that this is acceptable. If, unusually, the deviation is large, a critical examination of the various tasks and associated penalty scores is carried out to determine the areas of possible error, and amendments are introduced.

7.4 Evaluation procedure

An AEM evaluation procedure is shown in Fig. 7.3. Stage 1 of the procedure involves preparatory work in identifying the various tasks to be analysed. In stage 2, part-attaching sequences and attaching operations are matched to the elementary task symbols. In stage 3 the evaluation indices are calculated, and in stage 4 a judgement is made as to the effectiveness of the design. If the design is considered to be unacceptable because either the evaluation score ratio, E, for a particular part is less than 80 or the assembly cost ratio, K, is greater

Step / Examples	Product structure and assembly operations		E_i: Part assemblability evaluation score	E: Assemblability evaluation score	K: Assembly cost ratio	Part to be improved
Structure 1 (before improvement)	C(↓↻) B(↓···) A(↓ −)	1 Set chassis A	100	73	1	B
		2 Bring down block B and hold it to maintain its orientation	50			
		3 Fasten screw C	65			
Structure 2	C(↓↻) B(↓) A(↓ −)	1 Set chassis A	100	88	approx. 0.8	C
		2 Bring down block B (orientation is maintained by spot-facing)	100			
		3 Fasten screw C	65			
Structure 3	B(↓···) A(↓ −)	1 Set chasis A	100	89	approx. 0.5	B
		2 Bring down and pressfit block B	80			

Figure 7.3 Assemblability evaluation and improvement examples

than 0.7, design improvement commences. The AEM methodology contains step-by-step design improvement tools which make specific, pertinent suggestions as to how the design may be improved by offering design improvement examples.

An illustration of a simple redesign procedure is shown in Fig. 7.3. Here, it is necessary to attach a small block, B, to a chassis, A, and the initial method, shown as Structure 1, involves the use of a bolt, C. In this method the block needs to be held during the screwing operation to maintain its orientation and this results in two poor assembly evaluation scores, one for the block because it needs holding down and one for the bolt because it is a threaded fastener operation. This results in a product assemblability evaluation score of 73 which is less than the acceptable value of 80. Examining the data, the holding down to maintain orientation is the worst individual evaluation score and the suggestion is that the need for holding is removed by spot-facing the chassis shown as Structure 2. This gives an improved evaluation score for the task of 100 and an improved evaluation score for the product of 88; the cost ratio as a result of this exercise improves to 0.8. In Structure 3, the bolt has been removed and the block attached to the chassis by using a press fit. The assembly evaluation score for the press fit is less than that for simple block placement and reduces from 100 to 80 but, importantly, one part has been eliminated. As a consequence, although the product evaluation score has not significantly improved (89 compared with 88), the assembly

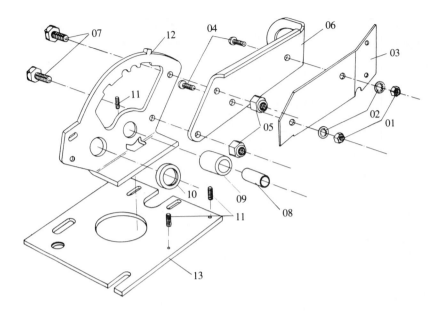

Figure 7.4 Original design of base sub-assembly

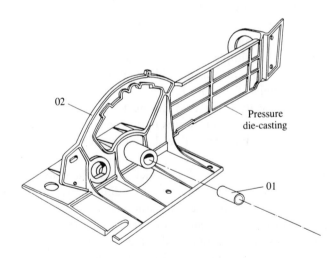

Figure 7.5 Redesign of base sub-assembly

cost ratio has significantly improved because of the reduced number of parts.

AEM presents clearly the effects of the design quality improvements, the difficulty of operations and the effect of reducing the number of parts.

Figure 7.4 shows a complex product, and the results of redesign using

AEM are shown in Fig. 7.5. As a result of the study, the total number of parts reduced from 18 to 2. The other significant factor which this and many other studies indicate is that when designs are extended to give improved or different functionality, this invariably results in fabrication which in turn often leads to the use of many separate fasteners.

CHAPTER 8

The Lucas DFA Evaluation Method

The Lucas design-for-assembly method arose out of collaborative work between the Lucas Organization and the University of Hull. The first commercial computer-based version was launched in October 1989, following a period of successful use of a paper-based version. The method is based around an 'assembly sequence flowchart' (ASF). The Lucas/Hull group has developed a knowledge-based evaluation technique, the Lucas DFA Evaluation Method, that systematically follows a procedure in which the important aspects of assemblability and component manufacture are considered and rated. The system is meant to be integrated into a CAD system, and because of this it should be possible to obtain most of the information required for the analysis work with the minimum of effort and time; this represents a major advantage over most other systems that effectively operate in stand-alone mode. The architecture of the whole knowledge-based system is shown in Fig. 8.1. The strategy and detailed rules of preceding sections are implemented using the logic programming language Prolog, and are designed to investigate a collection of Prolog clauses (rules) to the satisfaction of system conclusions using backward chaining. The system knowledge is transparent and is displayed, accessed and modified readily; the system interpreter can also provide explanations of decision-making activities.

8.1 Evaluation procedure

Evaluation of the product design is carried out using the procedures shown in Fig. 8.2. As product design commences, it is important to decide whether the product is unique or whether there are similarities, and therefore opportunities, for standardization of components and/or assembly procedures, and the establishment of a product family theme.

112

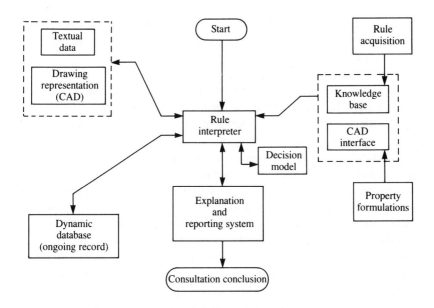

Figure 8.1 Architecture of the Lucas DFA (design for assembly) system

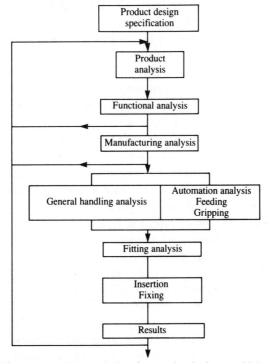

Figure 8.2 The Lucas DFMA (design for mechanical assembly) procedure

This is a further feature which is not evident in most other DFA systems; most consider only the current assembly and do not access prior knowledge for 'guidance'.

Functional analysis is carried out according to the rules of value analysis and activities are categorized by degree of functional importance. Each activity (addition of part or operation) is inputted to the system in turn with a name, a description and, for parts, a short sequence of questions about relative motion, special material properties and assemblability issues, the answers to which are used to determine the requirement for the part to be separate from all other parts already assembled. Each activity is then assigned to be either an essential (category A) activity or a non-essential (category B) activity, and the information is recorded on the assembly sequence flow-chart. The design efficiency is then defined as the ratio of category A activities to all activities $(A/(A + B))$ and at this stage, if the design efficiency is low, a redesign might be prompted before a detailed analysis is carried out; a suggested design efficiency threshold is 60 per cent. Assessing the design at this point is extremely useful since it causes interaction with other important manufacturing and marketing criteria at an early stage in the design exercise. It is pointless to carry out a detailed study only to find at a later stage that much needs to be changed because of conflicts.

Assembly costs can be reduced by the elimination or combination of components (or operations) falling into category B. A consequence of this type of functional analysis is that it is possible to best accommodate the problems of assembly by designing with a few complex parts as opposed to many simple parts; this might compromise other manufacturing requirements. Because of this, a complementary manufacturing analysis has been developed, the goal of which is to enable the designer to anticipate the consequences of assembly design rationalization on manufacturing costs. The manufacturing analysis allows the exploration of alternative materials and manufacturing technologies, and enables a cost value to be assigned to component design alternatives. The manufacturing analysis procedure, using data readily available to the designer, has been developed both as a stand-alone tool and as an integral part of the DFA evaluation procedure.

The Lucas/Hull method differentiates between manual and automated assembly by using the term 'handling' when components are dealt with manually and 'feeding' when components are handled automatically. For situations where automated assembly is appropriate, the evaluation procedures concentrate primarily on automatic parts feeding; conversely, for manual assembly, the evaluation favours the insertion operations.

This is consistent with the technology; automated assembly is almost exclusively dependent on the ability to automatically feed and orient small parts, whereas in manual assembly, handling parts is usually simple and it is the insertion of parts which has a higher potential for being problematic.

The feeding (or handling) analysis consists of questions about each component, the answers to which are used to determine a feeding index. For the automated feeding analysis, the methodology will guide the user towards a preferred feeding technology. An example taken from the knowledge base is given below:

Example: Substrate cover—*if* component is a flat regular prism *and* can maintain a stable orientation on track *and* has only slight asymmetry *or* has features too small for mechanical tooling *then* possible method
orient with remote (optical, laser) tooling
cost index 3
design advice: consult FRP file

It is also useful to determine the orienting efficiencies for the device (the ratio of feeding device output to feeding device input) and to have, an idea at least, of how much the feeding is likely to cost (how complex the feeder tooling is and how long it will take to make it).

An example of an orientation rule is:

Example: *If* component has rotational symmetry *and* concealed end-to-end asymmetry *and* mass not at geometric centre *and* part envelope is a long cylinder *then* possible method
orientate using swing tooling
efficiency 0.5
tooling time 15 hours

The next stage in the evaluation process is to determine whether the part has any characteristics that will cause problems in parts conveying. Examples of these are tangling, nesting, fragility, etc. A typical menu item for this is shown in Fig. 8.3, which shows a specific question and its 'help' information. After these questions have been answered, the part is assigned a feeding index and it is suggested that this should be less than 1.5; if this is not the case, the designer's attention is drawn to this and suggestions are made as to how the feeding index might be improved.

After establishing a feeding index for all parts, the feeding ratio is determined and this is defined as the sum of *all* the feeding indices divided by the total number of *essential* parts. The feeding ratio gives a good indication of the overall suitability of a design for automatic parts

---

Handling Analysis of 4 Bolt Gearbox
Question: Does the part tangle?
Explanation: Does the part have open features or projections which cause tangling when in bulk, and will require some manipulation to be separated out? e.g. circlips, some wire and pressed components, coil springs where the wire diameter is less than the free helix spacing.
Press Enter to continue

Figure 8.3 Handling analysis questions determining ease of handling

handling; a reasonable value of feeding ratio would be no greater than 2.5. It is suggested that if the feeding ratio is greater than 2.5, efforts should be made to reduce this by concentrating on parts that still have a feeding index of greater than 1.5 and by re-evaluating the decisions that led to the use of non-essential parts. While the feeding index serves no specific purpose for a particular product, it is a very useful comparison tool when considering alternative designs.

8.2 Assembly sequence flow-chart

After completing the feeding analysis, the user proceeds with the fitting (insertion) analysis. The fitting analysis requires the designer to generate an assembly sequence flow-chart and to assign a fitting index to each of the individual processes. The objective of this analysis is to identify fitting processes that are expensive and, more importantly, to give indicators as to how these processes might be changed so as to reduce their cost. The basic processes are relevant to manual assembly; if automated assembly is to be considered, a supplementary set of questions is asked. The additional questions relate to automatic part orientation and gripping, and are not relevant to manual assembly. The Lucas/Hull method considers the following processes:

- Work-holding—placing a temporary part to act as a fitting aid such as a guide, a holding-down device, a spacer, etc.

- Inserting and fixing—alignment, clearances, positioning requirements, tolerances for insertion operations; fastener type, fastener condition, etc. for fastening devices.

- Non-assembly operations—adjustments, re-orientations, calibrations, inspections, etc.

- Orienting and gripping (automated assembly only)—degree of manipulation, type of grip, size of grip, etc.

For all these points, questions common to all fitting tasks are also asked, such as what is the direction of fitting, complexity of fitting (compound or simple motions), etc?

Like other good DFA methodologies, the Lucas/Hull system attempts to identify, in order of frequency of occurrence, the various elements of fixing such that effort is concentrated on those activities which occur most often and which, consequently, have the most significance.

As for feeding, individual activities are assigned a fitting index for which a perfect activity generates an index of 1. It is suggested that activities with a value greater than 1.5 should be reviewed with a view to improving the design. A typical menu item with its associated 'help' information is shown in Fig. 8.4. The fitting ratio is defined as the sum of *all* the individual fitting indices divided by the total number of *essential* parts. A recommended maximum value for this is 2.5 with a recommendation again that, for values greater than this, individual activities should be examined as should the non-essential parts.

Figure 8.5 shows a typical example of an assembly sequence flow-chart for a manually assembled pump. It can be seen from this example that there is a very large number of non-essential parts in the original design and, not unexpectedly, most of these are separate fasteners. The feeding indices are generally good but, of course, with so many non-essential parts involved, the sum of the feeding indices is large when compared with the number of essential parts. A similar trend can be seen for the fitting indices; many of the individual values are acceptable but, again, the values ascribed to non-essential parts completely overwhelm those for essential parts, and the resulting fitting ratio is completely unacceptable.

The redesigned pump is shown in Fig. 8.6 together with the new assembly sequence flow-chart, where it can be seen that all the separate fasteners have been eliminated by the use of integral fastener snap-fits, the valving has been simplified and use has been made of a

---

F2 = rule
F3 = stop analysis
Explanation --
Alignment and position of the part during insertion can be made difficult if
    there are problems such as:
    No lead or entry chamfers on parts.
    No lead-in shank on threaded parts or pilot diameters on threaded holes.
    Closely toleranced holes and shafts that tend to snag.
    A definite orientation is required, e.g. inserting a shaft with splines.

Figure 8.4 Question and explanation in insertion analysis routine

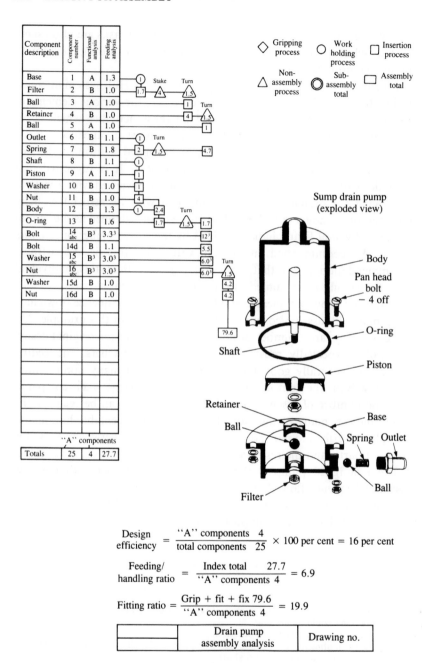

Component description	Component number	Functional analysis	Feeding analysis
Base	1	A	1.3
Filter	2	B	1.0
Ball	3	A	1.0
Retainer	4	B	1.0
Ball	5	A	1.0
Outlet	6	B	1.1
Spring	7	B	1.8
Shaft	8	B	1.1
Piston	9	A	1.1
Washer	10	B	1.0
Nut	11	B	1.0
Body	12	B	1.3
O-ring	13	B	1.6
Bolt	14 abc	B³	3.3³
Bolt	14d	B	1.1
Washer	15 abc	B³	3.0³
Nut	16 abc	B³	3.0³
Washer	15d	B	1.0
Nut	16d	B	1.0
		"A" components	
Totals	25	4	27.7

◇ Gripping process   ○ Work holding process   ▢ Insertion process

△ Non-assembly process   ◉ Sub-assembly total   ▢ Assembly total

Sump drain pump
(exploded view)

$$\text{Design efficiency} = \frac{\text{"A" components}}{\text{total components}} \frac{4}{25} \times 100 \text{ per cent} = 16 \text{ per cent}$$

$$\text{Feeding/ handling ratio} = \frac{\text{Index total}}{\text{"A" components}} \frac{27.7}{4} = 6.9$$

$$\text{Fitting ratio} = \frac{\text{Grip + fit + fix}}{\text{"A" components}} \frac{79.6}{4} = 19.9$$

	Drain pump assembly analysis	Drawing no.

Figure 8.5 Lucas method—assembly sequence flow-chart example

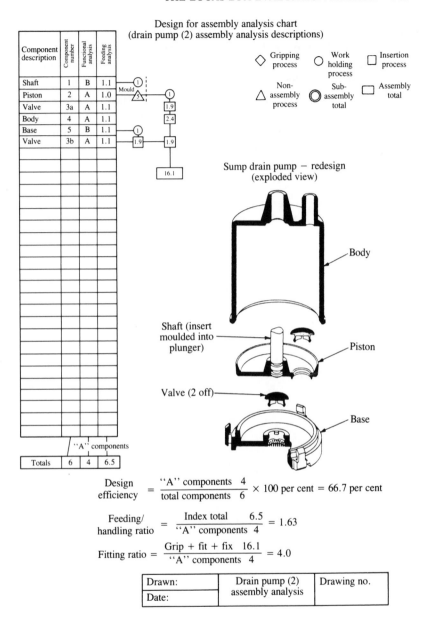

Design for assembly analysis chart
(drain pump (2) assembly analysis descriptions)

Component description	Component number	Functional analysis	Feeding analysis
Shaft	1	B	1.1
Piston	2	A	1.0
Valve	3a	A	1.1
Body	4	A	1.1
Base	5	B	1.1
Valve	3b	A	1.1
"A" components			
Totals	6	4	6.5

◇ Gripping process   ○ Work holding process   ▢ Insertion process

△ Non-assembly process   ◎ Sub-assembly total   ▭ Assembly total

Sump drain pump — redesign
(exploded view)

Body

Shaft (insert moulded into plunger)

Piston

Valve (2 off)

Base

$$\text{Design efficiency} = \frac{\text{"A" components}}{\text{total components}} \frac{4}{6} \times 100 \text{ per cent} = 66.7 \text{ per cent}$$

$$\frac{\text{Feeding/}}{\text{handling ratio}} = \frac{\text{Index total}}{\text{"A" components}} \frac{6.5}{4} = 1.63$$

$$\text{Fitting ratio} = \frac{\text{Grip + fit + fix}}{\text{"A" components}} \frac{16.1}{4} = 4.0$$

Drawn:	Drain pump (2) assembly analysis	Drawing no.
Date:		

Figure 8.6 Lucas method—redesigned example

suitable piston material to eliminate the O-ring. These and other small modifications lead to a reduction in non-essential parts from 21 to 2 with the subsequent improvements in both the feeding and fitting ratios. As could be expected, the average feeding and fitting ratios (sum of indices divided by total number of parts) have hardly changed (1.11 to 1.15 and 3.18 to 2.7 respectively). This reinforces the view expressed throughout that by far the most significant element of redesign for assembly is the attempt to minimize the number of parts.

CHAPTER 9

The Boothroyd–Dewhurst DFA Method

Boothroyd–Dewhurst Inc. have now developed design-for-manufacture as well as DFA methodologies; however, only their DFA methodology will be discussed here. Product design exercises can be carried out using either their handbook or their DFA software. Here, the general principles behind the methods will be discussed with reference to the handbook.

9.1 Objectives of the method

The Boothroyd–Dewhurst DFA method addresses the problems of:

- determining the appropriate assembly method,

- reducing the number of individual parts that must be assembled, and

- ensuring that the remaining parts are easy to assemble.

The first step in their procedures is to select the appropriate assembly method for the product. The designer must decide, from the values of the basic product and company parameters (number of parts, production volume, etc.), which assembly method is likely to be the most economic. The ways of assessing the available assembly methods are summarized in a chart, a cutaway of which is shown in Fig. 9.1. The chart is based on an analysis of mathematical models of the various assembly processes. The methods of assembly are classified into three basic categories.

1 Manual assembly—bench or transfer-line assembly using only simple tools.

2 Special-purpose transfer machine assembly—assemblies are transferred by an indexing transfer device (rotary or in-line).

3 Robot assembly:

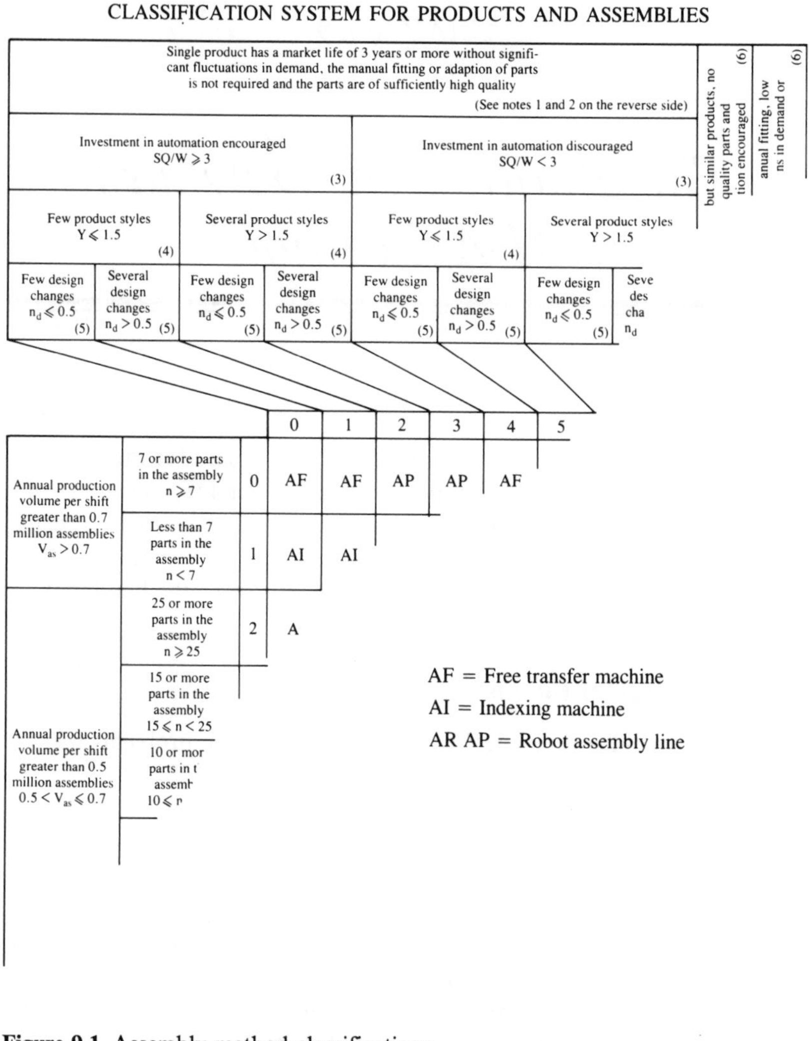

CLASSIFICATION SYSTEM FOR PRODUCTS AND ASSEMBLIES

				but similar products, no quality parts and tion encouraged (6)	anual fitting, low ns in demand or (6)
Single product has a market life of 3 years or more without significant fluctuations in demand, the manual fitting or adaption of parts is not required and the parts are of sufficiently high quality (See notes 1 and 2 on the reverse side)					
Investment in automation encouraged SQ/W $\geqslant$ 3 (3)		Investment in automation discouraged SQ/W $<$ 3 (3)			
Few product styles Y $\leqslant$ 1.5 (4)	Several product styles Y $>$ 1.5 (4)	Few product styles Y $\leqslant$ 1.5 (4)	Several product styles Y $>$ 1.5		
Few design changes $n_d \leqslant 0.5$ (5)	Several design changes $n_d > 0.5$ (5)	Few design changes $n_d \leqslant 0.5$ (5)	Several design changes $n_d > 0.5$ (5)	Few design changes $n_d \leqslant 0.5$ (5)	Several design changes $n_d > 0.5$ (5)	Few design changes $n_d \leqslant 0.5$ (5)	Seve des cha n_d

			0	1	2	3	4	5
Annual production volume per shift greater than 0.7 million assemblies $V_{as} > 0.7$	7 or more parts in the assembly $n \geqslant 7$	0	AF	AF	AP	AP	AF	
	Less than 7 parts in the assembly $n < 7$	1	AI	AI				
	25 or more parts in the assembly $n \geqslant 25$	2	A					
Annual production volume per shift greater than 0.5 million assemblies $0.5 < V_{as} \leqslant 0.7$	15 or more parts in the assembly $15 \leqslant n < 25$							
	10 or mor parts in t assemb̶ $10 \leqslant n$							

AF = Free transfer machine
AI = Indexing machine
AR AP = Robot assembly line

Figure 9.1 Assembly method classifications

(a) one general-purpose robot arm operating at a single workstation,
(b) two general-purpose robot arms work hand-in-hand at a single station,
(c) a multi-station free-transfer machine with general-purpose robot arms.

Once it has been decided that the product or assembly is likely to be assembled manually, by special-purpose assembly machine or by robot assembly, then an analysis of the design, identifying the assembly difficulties and estimating assembly costs, is made for the chosen assembly method.

The most powerful tool of this or any other design-for-assembly system is the reduction in the number of parts required for the product to be functionally acceptable. This was identified in Chapter 3 where it was shown that, if a different material is necessary, if there is relative motion or if the ability to physically perform the assembly function is impaired, separate parts are needed.

The secondary but extremely effective tools of the system are based on a system of penalties which for the particular activity, be it manual or automatic handling, manual, automatic or robot insertion, poses the question: 'How does what I am attempting to do, compare with how it could be done if every aspect of the part design favoured the activity?' This skilfully covers two important elements of the design activity. It allows meaningful quantitative judgements to be made and, very importantly, it gives the user the opportunity to view easily the redesign options available. The importance of this cannot be overstressed. Knowing that a design is less than adequate is useful but, without advice on what can be done about it, the knowledge can never be fully utilized.

With this strategy in mind the Boothroyd–Dewhurst system then clarifies parts for the particular assembly activities and assigns to the classifications the various penalties as described above. The first judgement to be made when classifying is, how many categories? Clearly, a system with few categories would be easy to use but not very useful, while a system with many categories would be useful but difficult to use. The Boothroyd–Dewhurst system very effectively uses the middle ground by firstly posing the questions:

• Which independent parameters affect the activity?

• How can these be graded in order of effect?

• How many need to be considered to cover *most* applications?

Since, as with most real activities conditioned by independent parameters, a law of diminishing returns applies, it is unusual for all parameters to have equal effects; a judgement can be made to limit the parameters to a usable number without significantly affecting the reliability of the system. The basic operation of the system will now be described by reference to a simple product.

9.2   Redesign of a simple product

Figure 9.2 shows a simple sub-assembly used in the construction of a gas-flow meter. The objective is to analyse the design using the Boothroyd–Dewhurst method with the intention of using the information obtained to create a new, easier-to-assemble, less expensive sub-assembly. While the Boothroyd–Dewhurst method can be used for product design for manual, dedicated and flexible assembly, for the purposes of brevity, only manual assembly will be considered in this analysis.

As is common for the redesign of an existing product, it will be assumed that the functional parts must have the same dimensions and be made of the same materials. Applying the simple rules of potential redundancy, it is clear that the product needs only one part if the trouble were taken to produce the part by complex metal removal processes. Clearly, this would not be an acceptable economic solution and a less optimal but more pragmatic design is needed. The obvious first level of relaxation is to use only two parts, the plate and the bearing housing (all the other parts are merely fasteners whose only function is to hold the functional parts in the required spatial relationship). These could be

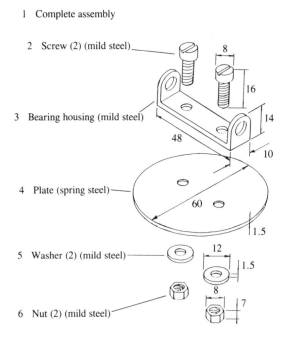

1   Complete assembly

2   Screw (2) (mild steel)

3   Bearing housing (mild steel)

4   Plate (spring steel)

5   Washer (2) (mild steel)

6   Nut (2) (mild steel)

Figure 9.2 Small consumer product sub-assembly

attached to each other by designing two pseudo-bifurcated rivets into the bearing housing (Fig. 9.3) such that these can be swaged over after insertion through the holes in the bearing plate. This would require specialized manufacturing equipment which might or might not be cost-effective.

Having successfully driven the designer towards a minimum part solution, the Boothroyd–Dewhurst system now focuses the designer's thoughts on what is wrong with the parts as regards handling and insertion. Assuming that this is to be manual assembly, pertinent questions are asked about handling (grasping, manipulating, etc.). Both parts are easy to grasp but they do pose manipulation problems; the plate can be assembled either way up but in a specific rotational orientation. The bearing housing can only be assembled one way up in a specific rotational orientation. Is there anything that can be done about this?

An obvious solution is shown in Fig. 9.4, where the rotational

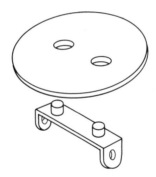

Figure 9.3 Redesign solution using an integral fastener solution

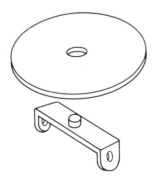

Figure 9.4 Redesign solution using more symmetry

asymmetry of the plate has been removed and the bearing housing has been modified to suit. The ability of the plate/bearing housing combination to resist a torque has now been severely impaired and, again, reference to the specification is needed.

Simplistically, the assembly costs will have been reduced to less than 25 per cent of the original cost (two parts versus eight parts). More importantly, however, the system has successfully forced the designer to consider all the important aspects of design for assembly and in consultation with others, to make reasoned judgements.

The whole of the above procedure was done with limited detailed knowledge of design for assembly—this is appropriate for simple exercises and for people skilled in applying design for assembly. For more realistic evaluations, and particularly for designers new to design-for-assembly considerations, this approach is not appropriate; examining the problem in a structured way using either the handbook or the software is to be advised. It could be argued that the strength of the methodology is that it is thorough; this implies that short-cutting is difficult and the probability of not taking important information into account is small.

9.3 Using the handbook

This same example is now considered using the design-for-assembly handbook. Figure 9.5 shows a worksheet for the product shown in Fig. 9.2, where it can be seen that the various parts and operations necessary to assemble the product are listed in an order of assembly in column 1. Most real products have many different assembly sequences, and choosing the best for particular circumstances is important when minimizing assembly costs. However, experience has indicated that in design for assembly that is first and foremost a method of *comparing* alternative designs, any *reasonable* sequence is acceptable.

The worksheet is now completed by reference to the appropriate sections of the handbook and by filling in the boxes in each row (for each part) in turn. The following will describe the actions necessary to fill in the first row of the worksheet, which is for the two nuts.

Number of repeats of the activity

This is an expedient which, for manual assembly, merely reduces the number of rows on the worksheet. For dedicated assembly this is more significant since there is the potential for doing the tasks simultaneously, and in robot assembly, there is either the potential for doing them

Design for manual assembly worksheets

Part I.d. number	Number of repeats	Manual handling code	Manual handling time penalty	Manual insertion code	Manual insertion time penalty	Total assembly time	Manual assembly cost	Figure for estimation of theoretical minimum number of parts	Remarks
6	2	0 0	0	0 0	0	6	1.2	0	Nut
5	2	0 0	0	0 0	0	6	1.2	0	Washer
4	1	1 2	1	0 0	0	4	0.8	1	Plate
3	1	3 0	0	0 0	0	3	0.6	0	Bearing housing
2	2	0 0	0	4 8	7	20	4	0	Screw
1	–	–	–	–	–	–	–	–	Complete assembly
						39	7.8	1	

Design efficiency $= 3 \times$ min parts/assembly time $= 3 \times 1/39 = 0.077$

Figure 9.5 Original design

simultaneously or for reducing gripper changing, etc. For the nut in this assembly, the number is two and this is placed in column 2 of the worksheet.

Manual handling code

A cutaway of the manual handling code data sheet is shown in Fig. 9.6. This coding sheet, in common with all the others, consist of a nominally 10×10 matrix with the important manual handling data along each axis. The matrix has been construed so that 'better' is achieved by moving up the page and to the left, i.e. the various handling activities are listed in order of increasing difficulty (handling time). This is useful in that, not only does the method give a reasonable indication of handling cost and an even better indication of relative handling cost (one action compared with other actions using the same handling methods), it also gives a very good indication of the redesign possibilities. To obtain the handling code requires the user to move down the code sheet and then across the code sheet until a category is reached that best describes the part under consideration. For the manual handling code sheet it can be seen that these questions consider the problem of ease of manipulation, and categorize these into:

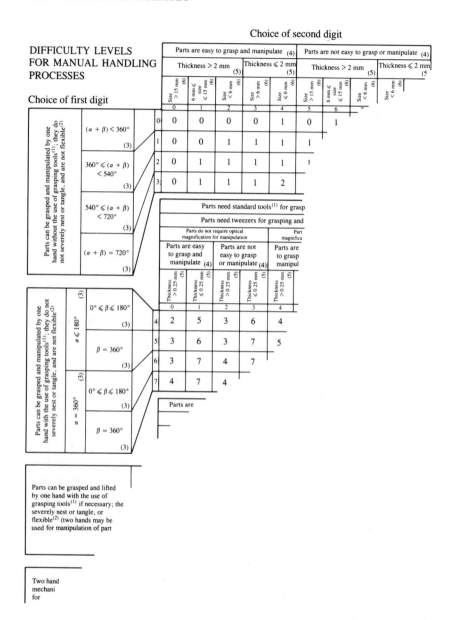

Figure 9.6 Manual handling code

- one-handed

- one-handed with grasping and manipulation aids

- two-handed because of manipulation

- two or four-handed because of size or mass

Because single-handed is by far the most prevalent manual handling activity, this is sub-categorized into four possibilities for each category of single-handed grasping; each sub-category effectively asks 'for a random orientation grasp, how far is the part from being the right way up and the right way round for insertion?' For the nut in this problem, it is one-handed, it can be any way up ($\alpha = 180$), it can, ostensibly be any way round ($\beta = 0$), it does not need grasping tools, it does not severely tangle or nest, and it is not flexible; this gives a first digit of zero.

The questions across the page can now be answered and these depend on the broad categorizations of the first digit response. For one-handed operations they are concerned with ease of grasping (not slippery, not delicate, etc.), and size or, more pertinently, lack of size; for one-handed operations with tools they are concerned with the type of tool and the ease of using the tool (how long it takes); for small parts which require two hands the main concern is the time to untangle or eliminate nesting; for two-handed or two-people activities where the problem is size or weight, the main penalties are due to manipulation (two hands) and additional labour (two people).

For the nut under consideration, it is easy to grasp, it is not delicate, fragile, etc., and it is not 'small'—the second digit, therefore, is zero. Thus the handling code is 00; this is entered in column 3 of the worksheet.

Manual handling time penalty
This figure is the penalty in seconds of any particular difficulty that might arise in grasping and manipulation, i.e. the simplest grasping and manipulation activity is defined as that which has none of the handling problems outlined on the coding sheet; to do this will take a fixed time. The figure in the 00 box on the coding sheet is 0 and this is entered in column 4 of the worksheet.

Manual insertion code
A cutaway of the manual insertion code sheet is shown in Fig. 9.7 where, using the same principles as for handling, the various insertion possibilities are considered. First, these are broadly categorized as:

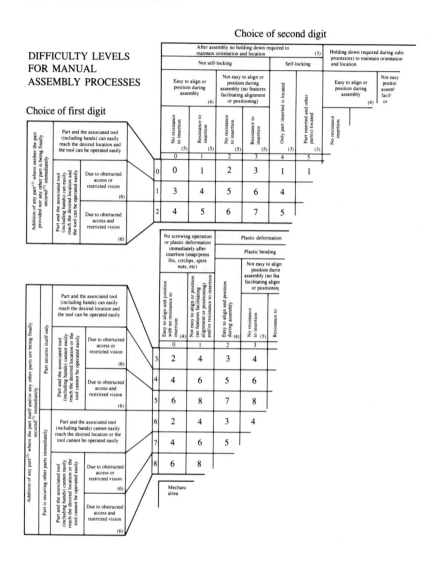

Figure 9.7 Manual insertion code

- parts placed and not secured

- parts added and secured (potentially extra time for securing operations)

- operations (activities that do not involve discrete parts)

Using the principle again that activities which are most common should be sub-categorized, all parts are subjected to the questions:

- Can the assembly location be seen?

- Is the assembly location accessible?

Again, as for the handling information, other pertinent questions are based on the broad categorizations already established. For non-secured parts, the questions involve 'holding down' (restricted capability for the operator), alignment and resistance to insertion. For 'fastened' parts, the type of fastening is important, and for operations, the type of operation will determine the time it takes to do it.

For the nut under consideration, it is not secured, and the assembly location can be both seen and accessed (it is being placed in the fixture); this gives a first digit of zero. It does not need holding down, it should be easy to align and position, and there should be no resistance to insertion (these are functions of fixture design); this gives a second digit of zero. The manual insertion code therefore is 00, which could be expected because this involves the relationship between a part and the fixture, and there is the potential for ensuring that the fixture best meets the requirements of the part. The insertion code figure is then placed in column 5 of the worksheet.

Manual insertion time penalty

Referring to the code sheet, for an insertion code of 00, the associated time penalty is 0. In the same way as for the handling times, an 'easy' insertion activity is defined and all more difficult activities receive a time penalty.

Total assembly time

With both the handling time penalty and the insertion time penalty, and the standard time for handling and insertion (three seconds), the total time to do the job can be obtained. In this case, because there are no penalties, the time for each nut is three seconds, giving a total time of six seconds; this figure is placed in column 7 of the worksheet.

Manual operation cost

This is merely a function of the assembly time and is based on the true cost of the labour; the value is placed in column 8 of the worksheet.

Minimum parts

The most significant result of a design-for-assembly study is that which gives a value for the minimum number of parts needed to achieve the necessary functionality. As mentioned previously, it is based on materials, motion and assemblability. For the part in question, the nut(s), this naturally falls into the category of separate fasteners which by definition are potentially redundant. For these, therefore, a zero is placed in column 9 of the worksheet.

The strength of the system now becomes apparent since an effective system should now indicate what should be done to produce a better design. From the full worksheet in Fig. 9.5 it can be seen that the total minimum parts count is one but, as indicated earlier, this would be rejected on cost grounds. If at least two parts are now necessary, these would have to be the bearing housing and the plate; these are both functional and all the others are merely fasteners. There are many possibilities for joining the bearing housing to the plate using integral fastening—one proposed solution is by the use of integral rivets as shown in Fig. 9.3. The worksheet for this solution is shown in Fig. 9.8 where emphasis is now placed on looking for further savings based on easier handling and insertion of the parts left. The plate can be placed either way up, but it does have rotational asymmetry and is 'thin'; one solution, as proposed earlier, is to have one axi-symmetric integral fastener as shown in Fig. 9.4. For this solution, the bearing housing cannot be improved since it still needs to be assembled one way up, one of two ways round, but the plate is now easier to handle; the worksheet for this solution is shown in Fig. 9.9. In this particular example there is now no penalty for the plate since it is still considered to be 'easy' to feed.

9.4 Conclusion

It is clear even for this trivial example that there are many possible solutions to any assembly problem and that the most appropriate depends on the production conditions such as production volume, life expectancy, available equipment, product market life, etc. However, a methodology which can successfully identify good designs for assembly with the potential for critically examining less optimal solutions must have considerable merit.

Design for manual assembly worksheet

Part I.d. number	Number of repeats	Manual handling code	Manual handling time penalty	Manual insertion code	Manual insertion time penalty	Total assembly time	Manual assembly cost	Figure for estimation of theoretical minimum number of parts	Remarks
3	1	3 0	0	0 0	0	3	0.6	1	Bearing housing
2	1	1 2	1	0 0	0	4	0.8	0	Plate
1	–	–	–	–	–	–	–	–	Complete assembly
						7	1.4	1	

Design efficiency = 3 × min parts/assembly time = 3 × 1/7 = 0.428

Figure 9.8 Redesign to minimize parts

Design for manual assembly worksheet

Part I.d. number	Number of repeats	Manual handling code	Manual handling time penalty	Manual insertion code	Manual insertion time penalty	Total assembly time	Manual assembly cost	Figure for estimation of theoretical minimum number of parts	Remarks
3	1	3 0	0	0 0	0	3	0.6	1	Bearing housing
2	1	0 2	0	0 0	0	3	0.6	0	Plate
1	–	–	–	–	–	–	–	–	Complete assembly
						6	1.2	1	

Design efficiency = 3 × min parts/assembly time = 3 × 1/6 = 0.5

Figure 9.9 Redesign to reduce handling and insertion difficulties

At each stage in the design process, a judgement has to be made on the economic implications of the proposed solution; this is perhaps where any design-for-assembly system fails and why Boothroyd and Dewhurst are currently addressing the problem of developing a complete design for manufacture package. In the authors' experience, design-for-assembly implementation works best using a team approach in which all involved parties are represented and whereby judgements can be made and compromise struck with minimum delay and minimum iterations.

PART IV

Design for Assembly and Conceptual Design

CHAPTER 10

Alternative Product Design Concepts

10.1 Introduction

The summaries of the various DFA methodologies given in Chapter 6 indicate that there are many different approaches to design for assembly but that there is essentially a high degree of commonality between the various methods and that any method, even if not specifically appropriate for a particular problem, is likely to prove beneficial. It could be argued that the strength of any DFA method lies in the structured approach that has to be adopted to make the method work. It can be inferred from this that the content of the method is of less importance, on the assumption that all the significant parameters are present, and that any advice on which method to adopt is more concerned with ease of use rather than meaningful philosophical arguments. Thus, for example, computer-based systems are invariably superior to manual systems because they are easier to use and less prone to errors, even though they do not usually offer more information to the user.

The importance and effectiveness of design for assembly is well documented for specific examples, and many companies have recognized just how important and effective it can be. What are not often appreciated are the less obvious benefits of using DFA methods. These, probably more important than the direct savings, can be identified as:

- Giving the designer insight into the philosophy—it has been stressed already that assembly is probably the designer's weakest discipline; this is sure to strengthen, given a familiarity with a DFA method. It has been suggested with justification that once a formal system has been used for a short time, the designer can create meaningful design without reference to the system.

- Encouraging standardization—many companies recognize that, in the past, they have given the designer too much licence in the selection of parts, particularly proprietary parts such as fasteners, seals, bearings, etc., and that considerable savings can be made by limiting choice.

- Encouraging commonality—in the authors' opinion, the most important bonus of design for assembly has been that of promoting commonality, particularly of sub-assemblies, so that a few sub-assemblies can be used in many products. The implications of this are far-reaching and the resulting savings probably far outweigh any specific savings achieved as a result of a design-for-assembly exercise. A subset of this has been the efforts made to ensure that products are customized, and costs incurred, as late as possible.

- Producing more reliable products—it is well established that reliability is a function of complexity and that one aspect of this is the number of parts involved. Therefore it follows that, by reducing parts, design for assembly results in improved reliability.

- Reducing manufacturing costs—most design-for-assembly proponents suggest that invariably a DFA exercise results in reduced manufacturing costs; there is now sufficient evidence to support this view.

- Reducing overheads—administering parts, whether they are being manufactured, being purchased or being assembled, is expensive and the more there are, the more expensive it becomes and the more likely it is that mistakes will occur.

10.2 Potential improvements in DFA systems

Given that there is basically little to choose from between different systems and they all have considerable merit, if only in the less obvious benefits, what can be done in the future to improve further what is already a very useful design tool? The most quoted criticisms of any DFA system is that it is 'stand-alone', that much of the information required already exists in a database, usually resident in a CAD system, and that this results in significant duplication of effort. Undoubtedly, there would be major benefits arising from interfacing DFA systems into CAD databases, but as yet nobody appears to be interested in doing this. Unfortunately, interrogating an existing database would not be fully effective since a database only contains information about parts, with no reference to the spatial relationships between parts. This is yet another example of blinkered thinking dating from the time that CAD systems were first developed; they only addressed aspects of design concerned with parts manufacture instead of considering the wider aspects, which would include assembly.

A CAD-resident DFA system has much more significance than merely making a design for assembly more 'accessible' and easier to use. As far as the authors are aware, there is no system which links design for assembly with assembly system design, yet there is no good reason why this should not occur. There is no doubt that in the future, when flexible assembly systems make a significant impression on the assembly market, there will be a need for designing products to meet the abilities of systems, designing systems to reflect the requirements of products, and much iteration and cross-referencing. Attempts are already being made to do some of this but it is very difficult when the required information is both in an unsuitable form and restricted.

Perhaps the most meaningful improvement to be made in design for assembly would be applying it at the conceptual design stage. The methods currently used are invariably applied *after* the conceptual design phase, since they work on identified parts, shapes and spatial relationships. These are then attenuated by applying the 'rules' to give more assembly-friendly results. As already reported, this results in far-from-trivial benefits but it does not really address the problem of conceptual design in which, given a specification, the designer attempts to meet a list of often-conflicting requirements associated with functionality, reliability, primary manufacture, assembly aesthetics, etc.

If anything prevents truly innovative design, it is the past. It is not coincidental that many familiar products look very much the same now as they did 50 and sometimes 100 years ago. Their design has been very much conditioned by what is current practice and this tends to stifle both creativity and sensible commercial judgements. Of course, there are many good reasons for maintaining the status quo, including:

- confidence in the design

- consumer conservatism

- fully paid-up tooling cost

- time to market

However, this cannot explain why so little truly innovative design is carried out and why good design-for-manufacture rules cannot be applied to conceptual design, particularly when designing 'new' products.

A good example of how design has been influenced by later thinking is the electronics industry, which by virtue of its relatively recent emergence has taken note of undesirable characteristics of, say, typical mechanical designs. In this industry it has been recognized that standardization would be useful and, importantly, not expensive. This has resulted in 'packages' that have shapes which are invariably completely unrelated to their functionality but which are very much concerned with assemblability. Of course, electronics are passive in the mechanical sense and the problems are less severe, but what has evolved does show how the approach to design is conditioned by the state-of-the-art in the formative years of design. Unfortunately, conservatism then sets in, usually resulting in design methodologies that are quickly frozen, in turn stifling creativity of design.

None of the above really helps to suggest what can be done to influence conceptual design in changing circumstances. As has already been pointed out, for assembly there are many factors that need to be considered which are process and method dependent but, in the main, these are attenuating features more relevant to refinement of designs than to conceptual design. Examination of design-for-assembly studies reveals two important design points:

1 For all forms of assembly, the most significant design feature is the optimizing of the number of parts in the product subject to the other design constraints. Potential savings resulting from this will vary depending on the assembly method and the assembly process but regardless of these, at least one half and usually more of all savings will result from this one activity.

2 The second most important design feature is good precedence, in which there are very few interactions between the various parts of the assembly and whereby particular circumstances can be accommodated by choosing the most appropriate assembly sequence. The ability to create good precedences is a function of optimizing the number of parts in the product, in that the fewer the parts, the more chance there is that parts will interact—but, of course, the less need there is for more scope in assembly sequence.

The important conclusion from this is that first consideration should always be given to optimizing the number of parts.

For manual assembly, these are the only two features that really matter and, importantly, these can be applied to conceptual design in a systematic way.

For automated assembly, handling, automatic parts feeding, grasping and parts inserting all have varying significance according to the assembly process, but these are invariably aspects of detailed design which can be accommodated once the concepts have been validated. This is very important since it leaves the designer free to concentrate on conceptual design without being inhibited or concerned with process and method-dependent considerations. Further, if it is possible to determine the optimum number of parts for an existing design or product, then by inference it should be possible to predict this for a concept, i.e. given a specification, it should be possible to establish the number of different parts required to achieve minimum functionality; this can then be moderated to achieve the specification. This is standard procedure in the evolution of a design but seems to be often ignored because of the factors highlighted previously.

To illustrate this, two products will be considered, one an essentially new product and the other, an existing product looked at in a different way.

10.3 Design of a disposable valve for beer dispensing

To achieve economies in the transportation costs for liquid products such as beer it has been concluded that one possibility is to transport the liquid in a large disposable container (the cost of the container is such that it is less than the cost of returning a more expensive, depreciating container). A consequence of this is that whereas, with a returned container, the valve through which the container is filled and emptied would be sterilized and reused, with a supposedly disposable container there would possibly be a temptation to reuse the container, which could present hygiene problems. The solution to this problem was considered to be to make the valve itself disposable in the sense that once it is uncoupled from the liquid (beer) dispenser it will not reseal and, consequently, it will allow air to enter the container. In these circumstances, however, it would not be possible to store the product for any significant time and would, therefore, be impractical.

Examining the functionality, the requirement is for:

- A valve(s) that will open and shut for a limited number of times (say 200) and in which the opening is activated by coupling the valve assembly to either the filling equipment (liquid entering the container) or the dispensing equipment(liquid being removed from the contain-er). The characteristics of this action are shown in Fig. 10.1.

- A valve(s) that will open and shut for a limited number of times (say 200) and which will:
  —open when the valve assembly is coupled to the filling equipment (gaseous back pressure being applied to assist liquid filling),

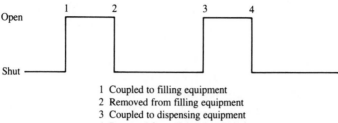

1 Coupled to filling equipment
2 Removed from filling equipment
3 Coupled to dispensing equipment
4 Removed from dispensing equipment

Figure 10.1 Characteristics of valve controlling liquid flow

—close when the valve assembly is uncoupled from the filling equipment (container filled),

—open when the valve assembly is coupled to the dispensing equipment (liquid being dispensed),

—remain open when the dispensing equipment is removed from the valve assembly (not reseal) or, alternatively, allow air to enter the container when the dispensing equipment is removed. This latter possibility was discounted since it implies piercing, significant forces, large motions and weakened material sections to a greater or lesser extent; i.e. leaving the valve open is only an extension of existing functionality, while piercing is additional functionality. The characteristics of the action necessary to leave the valve open are shown in Fig. 10.2.

• A valve assembly that will fit in the neck of the container (80 mm diameter) and be of an 'appropriate' length (< 75 mm).

• Valves that will adequately seal for the required number of operations and for the life of the product (say six months).

• Materials that will remain uncontaminated in normal transit and storage for the life of the product.

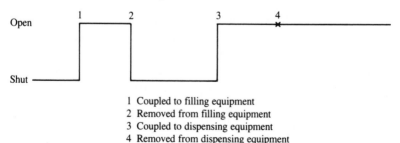

1 Coupled to filling equipment
2 Removed from filling equipment
3 Coupled to dispensing equipment
4 Removed from dispensing equipment

Figure 10.2 Characteristics of valve controlling gas flow

Additionally, the product should have a manufacturing cost of no more than 40 pence and a required annual production rate of one million per year.

Logically design for function can proceed as follows:

- The minimum number of parts necessary to give the *basic* functionality based on motion is three since there is a valve body, common to both valves, and two valves that need to move relative to the body to expedite opening and sealing.

- The minimum number of parts to give *basic* functionality based on different materials is not increased since there are no further parts that are necessary and hence no different materials. Conversely, if there is, subsequently, a reduced material cost, a reduced manufacturing cost or an improved functionality to be gained from using different materials for the various parts of the product then this should be considered, i.e. using the same material for parts that already need to be separate has no intrinsic merit.

- The obvious choices for material are stable metals such as stainless steel or copper/zinc copper/tin alloys but on the grounds of material cost, and to a lesser extent manufacturing cost, these would be unacceptable. An obvious non-metallic material would be some form of plastic which has a relatively lower material cost and a potentially lower manufacturing (moulding) cost.

- For the annual production volume required and using an appropriate manufacturing process, the question needs to be asked: can a suitable material be moulded such that a valve seat and matching valve can seal to conform to the specification? Clearly, this cannot readily be answered without suitable tests but it is important that this is resolved since the implications are significant; at a minimum, requiring, say, a different valve seat material would result in at least two additional parts and this would increase the number of parts by 66 per cent. This is a good example of the importance of initially establishing minimum functionality and building on this. Inferring what is appropriate is actually more expensive than observing what is specified.

- On the assumption that a plastic can be used, it is necessary to decide what characteristics this material should have and this will be determined primarily by the motion's specification. Choosing a suitable 'elastic' plastic is simple and integral elastic members can be designed in many forms; one obvious possibility is the use of suitably restrained struts as shown in Fig. 10.3.

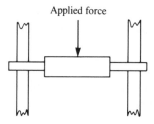

Figure 10.3 Motion with elastic characteristic

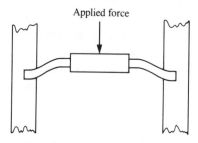

Figure 10.4 Motion with elastic characteristic followed by buckling

- The characteristics shown in Fig. 10.2 is less obvious in that the requirement is for an elastic characteristic up to a specified deflection followed by a permanent 'set' for larger deflections. Again using struts, an elastic/plastic strut could be designed which would 'fail' at a predetermined deflection; this is considered to be intrinsically unreliable since it would require high-quality, accurate section struts. A realistic alternative would be to use a 'buckling' strut which has elastic properties to a given deflection and which then buckles to give a much larger deflection as shown in Fig. 10.4. While this is not 'failure' in that the buckling can be reversed by motion in the opposite direction, if it is impossible to apply this motion then the valve would be permanently opened.

To summarize, at this point there are three parts that can be manufactured from the same or different plastics and for which the motions between the moving parts are as a result of deflecting struts. There is now a need to establish whether the necessary motions can be applied to the struts. On the basis that there are very many dispensing units and relatively few filling units, it is clear that modifying or discarding and replacing dispensing units would be much more expensive than modifying or replacing filling units, i.e. the motion elements should be designed to meet the dispensing specification and the filling units should be modified to meet the motion requirements for filling. In the event, the only modification required for the filling units is that the deflection which results from coupling should be such that the buckling strut is not deflected to the point of buckling.

It is not the purpose of this text to examine design for manufacture and, clearly, the next task in this exercise would be to examine manufacturing process possibilities and potential conflicts between these and functionality and assembly requirements. A possible design using a manufacturing process such as injection moulding is shown in the view of the three-part assembly shown in Fig. 10.5.

The example and the proposed solution are not important but the methodology used is significant. Too often, either a design is over-specified or the functionality is more than the specification requires, both of which result in additional cost.

10.4 A different approach to the design of a lever-arch file mechanism

The lever-arch file mechanism (Fig. 10.6) has been manufactured and successfully exploited for many years and is the standard 'quality'

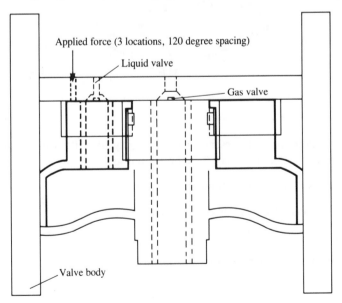

Figure 10.5 Valve arrangement to meet specification

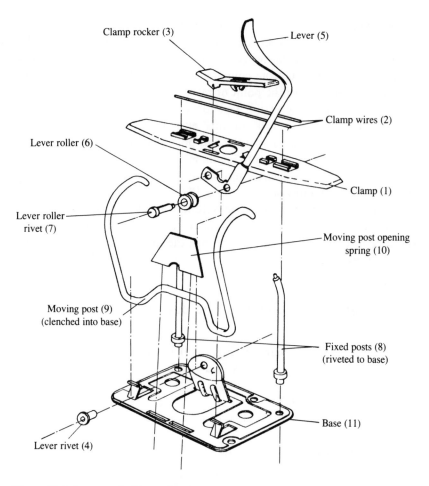

Figure 10.6 Lever-arch file mechanism

method of holding and preserving the integrity of written material. Its important characteristics may be summarized as:

- Robustness—it will withstand careless use
- Ease of inserting and removing pages
- Large storage capability
- Effective paper clamping for transportation
- Low cost

Lever-arch file mechanisms are manufactured in very large quantities by relatively few manufacturers. These manufacturers then sell the mechanisms to manufacturers of the lever-arch files who assembly the mechanisms to the file. Because the product has been in existence for many years, it reflects the state of technology of its time; it is of mainly metal construction and the majority of the manufacturing operations are metal forming. The equipment required to manufacture the parts is very specialized, and hence expensive, but the production volumes can easily justify the equipment cost. To successfully market a completely different yet functionally identical product would be virtually impossible because of customer resistance to change, initial capital equipment costs, advertising costs and subsequent potential for failure. It is an excellent example of a familiar consumer product which satisfies a customer need, which would be very difficult to displace and yet which in some ways is not consistent with the possibilities offered by modern manufacturing processes and methodologies.

Of course, a complete rethink of the design would almost certainly result in failure. Nevertheless, it makes an interesting exercise which indicates the transience of design and how the 'best' solution is, to a large extent, only a reflection of what was appropriate at a particular instant in time.

The processes used for manufacturing the parts of the mechanism would still be appropriate today if that was the only criterion. However, the mechanism has 12 parts and assembly is a significant cost item. Whereas 50 years ago labour costs could almost be ignored, today they are important and a different design would need to reflect this. If a conceptually different design is to evolve for an existing product, the specification needs to be examined carefully to ensure that it is correct.

The primary functional requirements are paper clamping, paper storing, mechanism opening and closing, mechanism clamping (closed) and paper transferring. Additional factors are cost, product life and size.

The paper-clamping facility is very important. When clamped, the file should be capable of being dropped without the clamp moving. If the clamp does move, there is a strong possibility that the paper would burst out and suffer damage. Equally importantly, the torn punched holes would need to be repaired and this would be a time-consuming exercise.

The feature which distinguishes the lever-arch file from other rigid file mechanisms is the quantity of paper that can be stored. To change this would destroy the product's identity and would be invalid.

Large storage capability and paper transferring imply constraining the paper over a distance with guidance and retention provided by the punched holes in the paper. It is difficult to envisage from the viewpoint of space anything other than a loop and from the viewpoint of guidance and retention anything other than 'wires'. If the product is to be both consumer and paper friendly, the wires must be smooth and the simplest (and potentially least expensive) smooth shape would be a 'wire' of circular cross-section.

Paper addition and removal requires that the loop be capable of being opened and that the position of opening is placed so that the design capacity for paper can be retained on both sides of the opening point. Further, while not mandatory, being able to remove or add paper with a uni-directional motion would be useful in that the operation could be done more quickly and with less potential for damage to the punched holes.

Mechanism clamping, while not vital, is useful for easy transferring of paper from one side of the file to the other. The clamping force, however, does not need to be high provided the paper clamp does its job effectively (the loop does not have to be closed for transporting).

The cost factor is obvious. The product is not for a restricted market and has to be within the means of many. There is little if no potential for upmarket variants and hence a low-cost unique product is appropriate.

The product life is perhaps contentious since life cannot be judged by a single criterion. One measure of life could be the number of times the file is opened and closed, another could be the number of times it is picked up and transported. A third could be the number of times it can be dropped, and finally the time from sale could be used. None of the above, alone, could be used in a specification but what is sure is that for the *majority* of applications the number of openings and closings is related to the capacity, in that once the file is full the frequency will fall dramatically—hence the number would, typically, be small (say 1000), the number of times it is transported would be small (less than the number of times it is opened and closed), the number of times it is dropped would be small (the ability to withstand being dropped once would be a reasonable criterion) and, for reasonable conditions, the shelf life would be high.

If the above is accepted and related to the current design, there is, to a degree, some over-design. The roller on the lever is strictly unnecessary; for the number of times the file is opened and closed, friction would not wear the contacting parts. The function of the spring could also be disputed; it merely serves to ensure that a positive force maintains the mechanism open when there is no real tendency for it to close. The camlock closing action is also more positive than is strictly necessary for paper transfer, and both the moving post and the fixed post are much stronger than is necessary, but this is a consequence of a preferred size and the material used.

Conversely, the paper clamp is not particularly positive. It relies exclusively on friction for its effectiveness, and personal experience indicates that it can move if the file is dropped. There are also too many parts in that one moving wire would be as effective as two.

When considering the other parts there is some potential redundancy. The lever, the lever arm, and the lever rivet could be made from a single piece since they do not move relative to each other, as could the two fixed posts; consideration of manufacturing costs would be needed. In theory, the base and the fixed posts could be integral but in this case even cursory consideration of manufacturing costs would indicate that this is inappropriate.

After addressing the problem of meeting the current specification, it is pertinent to look at any other features that a lever-arch file mechanism might have which would make it more attractive to the user. It has already been stated that a direct competitor would probably have little success but a mechanism with extra capability, particularly if there is a negligible cost penalty, might be competitive.

Two activities are synonymous with using a lever-arch file: punching paper and writing. For the former, a separate paper punch is used which is generally not particularly transportable. Further, for any one lever-arch file, the amount of paper that needs to be punched is small (typically the capacity of the file). If a punch could be incorporated into the design, it would not need to have the durability of a separate punch because it would not need to punch the same amount of paper, and it would also not need to be able to punch multiple sheets. For the latter, a facility for attaching a pen might be attractive; this is not as clear-cut as the punch since one pen is normally used to write for many files, but it would prove attractive to a sector of the market.

As for the previous example, the first thing to do is to establish motions and materials so that an estimate can be made of the minimum number of parts which would be required to give the necessary functionality. In this case there are two motions, that to allow opening and closing of the mechanism and that to allow paper clamping. Since both these motions are relative to a common part, there is potential for the one motion to perform both functions. Examination of this possibility led to the conclusion that it was not possible and for motion considerations, therefore, three parts are needed. For the additional functionality of punching, it was thought that the mechanism motion could be used; hence this would not increase the number of parts. For the pen-holding facility, it was thought that, with an appropriate material, a snap fit could be used and this also would not necessitate any more parts; these considerations led to the concepts shown in Fig. 10.7.

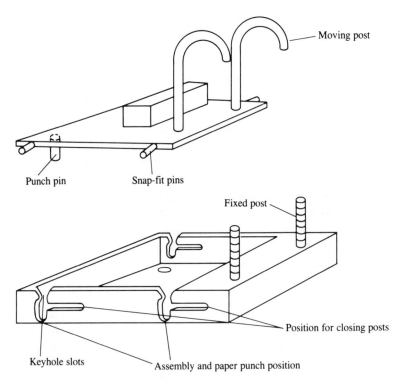

Figure 10.7 Conceptual lever-arch file

Turning to the paper clamp, it has already been said that this was not particularly effective. The fixed post could be profiled with grooves such that the clamp would engage positively in the fixed post. With sufficient grooves and some elasticity in the underside of the clamp, paper could be held securely; this is shown in Fig. 10.8.

The above example is perhaps extreme but it is included merely to illustrate that new concepts should receive the benefit of unbiased consideration which should only be invalidated when there are sound financial and commercial reasons.

Some more examples are now given to illustrate the basics of good conceptual design, formalized methodologies and related factors that can help the designer.

10.5 Developing alternative product concepts

The development of products should be carried out in accordance with a systematic design procedure. The design process is divided into four stages:

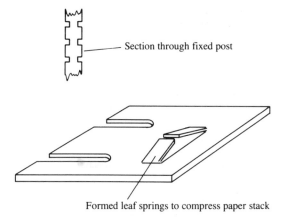

Section through fixed post

Formed leaf springs to compress paper stack

Figure 10.8 Paper clamp

1 Task clarification

2 Concept development

3 Product layout

4 Detailed design

Figure 10.9 shows the various stages of the process step by step. At each stage a decision has to be made as to whether to proceed or whether to backtrack and explore alternative possibilities, because of problems evident at the current point.

Clearly, manufacturing and assembly knowledge must be used effectively at all stages of design and must be embodied in the development of the product; this text will concentrate only on considerations for assembly. As pointed out earlier, during the various stages of the design process, different types of assembly knowledge need to be used and it is important not to 'over-design' at the early stages of design. The following builds on the thoughts expressed earlier as to what might be done in product design for assembly at the conceptual stage.

The conceptual design stage is that part of the design process in which identification of the essential problems takes place. This is done by the establishment of the specification, particularly for functionality, and by the search for appropriate solution possibilities and their potential combinations. The basic solution path is identified through the expansion of possible solution concepts. The main steps of conceptual design are:

1 Clarification of the specification

2 Interpretation of the specification

1. **Task clarification**
   1.1 Recognize the requirements
   1.2 Define the tasks
   1.3 Create a requirements list
   1.4 Release for conceptual design
2. **Concept development**
   2.1 Analyse functions
   2.2 Create the functional structure
   2.3 Create variations of the functional structure
   2.4 Determine solutions for each function
   2.5 Select principles
   2.6 Prepare solution variants
   2.7 Work out different concepts
   2.8 Evaluate and select the different concepts
3. **Product layout**
   3.1 Draft the main functional units
   3.2 Draft the remaining functional units
   3.3 Select appropriate parts of the draft
   3.4 Detail the design of the main and auxiliary functional units
   3.5 Check and improve drafts
   3.6 Analyse cost recovery
   3.7 Complete draft
   3.8 Make decisions regarding draft
4. **Detailed design**
   4.1 Detail the design
   4.2 Work out production specifications
   4.3 Analyse production data
   4.4 Release for production

Figure 10.9 The design process

3 Establishment of function structures

4 Search for solution principles

5 Selection of a suitable solution principle

6 Identification of concept variants

7 Evaluation of concept variants

Clarification of the specification

The specification must be defined as fully and clearly as possible so that amplification and correction during its subsequent elaboration can be confined to the most essential elements. An example is now given for the design of a water mixer tap. For this the specification is formulated as:

The development of a mixer tap as a device for the regulation of water flow and temperature as a one-handed device, with high reliability for stopping the flow.

The detailed functional requirements and other conditions are listed in a specification list which defines the functional requirements and constraints in detail.

Specification

Maximum pressure	15 bar
Maximum temperature	60°C (standard) 100°C (short time)
Mixed flow maximum	12 1/min at 2.5 bar
Service life	8 years
Number of variants	6

Interpretation of the specification

The specification is an input for a designer's analysis of the required functionality and constraints. The designer formulates the main problem as being:

The flow of hot water is either stopped or is so metered that the mixed temperature can be adjusted to any desired value regardless of the water flow rate. Further, on changing temperature, the water flow rate must remain unchanged.

The functions, therefore, are:

Meter
Stop
Adjust
Mix

Inputs are pressure, flow rate and temperature of hot and cold water; outputs are pressure, flow rate and temperature of mixed water. This data forms the basis for establishing the function structures.

Establishing the function structures

The function is the relationship between inputs and outputs of a system. The complexity of the problem results from the complexity of the overall function. A complex function can be divided into sub-functions of lower complexity; the combination of individual sub-functions results in a function structure. The objective of breaking down a complex structure is to establish more:

• possibilities in the search for solutions,

• possibilities for structuring the product as building blocks,

• open structures,

• effective product variant programmes,

- possibilities for varying variant and non-variant sub-assemblies,

- possibilities for varying function structures to avoid redundancy and for establishing a simple and unambiguous function structure.

Figure 10.10 shows one selected function structure for the mixer tap and this has been analysed; a structure with linear characteristics was selected.

Searching for solution principles to fulfil the sub-functions

A solution principle is a technical realization of functionality needed for the fulfilment of a function and its design features. There are many methods of searching; for example, conventional methods, methods with an intuitive bias and methods with a discursive bias can all be used. Which methods to use, in a particular case, depends on the problem, the access of information, etc. Searching for solution principles for the one-handed mixer tap was carried out using brainstorming methods. Figure 10.11 shows some solution principles generated to fulfil the required functional structure.

Combining solution principles to fulfil the overall functionality

To fulfil the overall functionality, it is necessary to combine solution principles. This means ensuring that there is physical and geometrical compatibility between the solution principles and the technical and economic constraints. For this purpose, morphological, matrix, mathematical and combinatorial methods have been developed.

Selecting suitable combinations

The designer must use a systematic and verifiable selection procedure which facilitates the choice of promising solutions. The selection

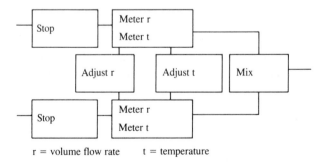

r = volume flow rate   t = temperature

Figure 10.10 A function structure for the mixer tap

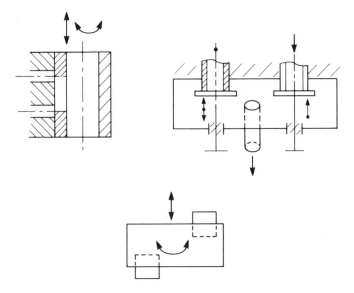

Figure 10.11 Solution principles (alternatives) of functional structure

procedure is based on the successful application of general criteria, which can be identified as needing to be:

- reasonable in principle

- compatible with the overall functionality

- low in cost, etc.

One method is the use of selection charts. Unsuitable principles can be eliminated and all others can be evaluated at the end of the conceptual design stage.

Identification of concept variants
The solution proposals are firmed up into concept variants. The concept variants must express the working principles and space requirements, and other task-specific constraints. Detailed information about the behaviour of concept variants can be obtained from:

- preliminary experiments

- preliminary models

- rough sketches

- rough calculations

From the viewpoint of assembly, important measures are:

- the number of parts
- applying new materials and technologies
- simple assembly sequences
- ease of inspection
- suitable spatial arrangement of parts

Evaluation of concept variants

Concept variants must be evaluated against technical and economic criteria. At this stage of design, there is not enough data for a precise cost analysis. Check-lists are useful for the evaluation of concept variants.

For the purpose of evaluating assemblability at the conceptual design stage, the following check-list can be used.

1 What is the design efficiency (based on a reasonable definition)?

2 Is the function structure one that will allow the suitable arrangement of parts (layered assembly, accessibility, etc.)?

3 Is it possible to consider modular design?

4 Has a base part in the product structure been identified?

5 Is the designed function structure such that a building block system can be applied?

6 Will it be possible to segregate variant and non-variant sub-assemblies?

7 Has the standardization of fasteners been considered?

8 Have standardized interfaces been planned?

9 Will it be possible to standardize checking, inspection, test, etc.?

10 Is it possible to use simple and inexpensive joining techniques?

In principle, any of the above questions receiving negative answers require those elements to be re-examined. Working through the check-list allows preliminary conceptual designs to be evaluated for assemblability. In this way, the requirements of assembly will be embodied in the product concept.

Figure 10.12 shows the consequences of selecting a good product concept on the main indicator of assemblability, the total number of

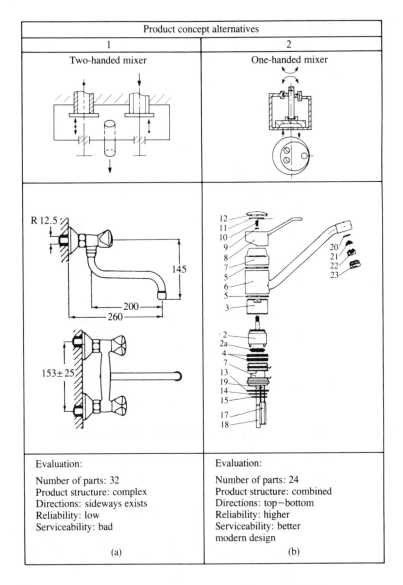

Product concept alternatives	
1	2
Two-handed mixer	One-handed mixer

Evaluation:

Number of parts: 32
Product structure: complex
Directions: sideways exists
Reliability: low
Serviceability: bad

(a)

Evaluation:

Number of parts: 24
Product structure: combined
Directions: top−bottom
Reliability: higher
Serviceability: better
modern design

(b)

Figure 10.12 Influence of product concepts on the assemblability

parts. In Fig. 10.12(a) is a product concept for a mixer with two valves which can achieve the required functionality. The solution in Fig. 10.12(b) is the result of a product concept which realizes the same functionality using only one valve; the two-valve mixer consists of 32 parts while the one-valve mixer has only 24 parts. Other assembly consequences of the design shown in Fig. 10.12(b) are that this product:

- is suitable for dedicated assembly,

- allows automated manufacturing of parts,

- has higher reliability,

- is easy to service,

- is a modern design.

From this design of product concept, it is possible to formulate one of the most important DFA principles:

> *Generate and analyse all solution principles which are able to use the latest science and technology concepts to fulfil the functional requirements. Prefer product concepts which lead to the minimum number of parts.*

DFA for conceptual design, as illustrated in Fig. 10.12 is, in principle, a combination of evolution and innovation. The overall functionality has been fulfilled by a more effective solution with a new arrangement of parts, but a totally new (revolutionary) product concept is missing. Genuinely 'new' products can only result from conceptual leaps which, unfortunately, do not occur very often. Nevertheless, significantly different and less expensive products can be developed by the application of sound design principles linked to lateral thinking.

10.6 Principles of assembly-oriented product structuring

The product concept resulting from conceptual design determines which individual parts will be in the product. The product concept is an input for the constructional design stage.

In constructional design, a designer must determine the structure of the product, i.e. the general arrangement and spatial compatibility of parts, the preliminary form designs, the dimensions, the functional surfaces, etc. Which product structure characteristics are relevant to assembly and how to change these is now dealt with in detail.

10.7 Product structure

Product structure refers to the number of elements (parts), their arrangement and the relationships between the elements. The properties of elements include shape, dimensions, material, surface quality and tolerance.

The design degrees of freedom which are suitable as regards ease of assembly will be dealt with in detail in the next section. Part design for assembly will not be considered; this has already been dealt with in previous chapters and by others at length.

The product structure is critical in that it determines the assembly method and the number and type of the assembly processes. Design practice recognizes two basic structures which strongly influence the number of parts: differential and integral.

Differential product structures

Differential product structures refer to the breakdown of a component into several parts. This contradicts the most important element of design for assembly, reduction in part count, but it is often necessary because of other considerations such as:

- the use of easily available and favourably priced semi-finished materials and standard parts,

- the use of larger numbers of simpler sub-assemblies with the inherent reductions in cost,

- a reduction in both the time taken and the cost of maintaining the product.

Obvious disadvantages of differential product structuring are:

- increased assembly costs,

- overall increased part manufacturing costs,

- the need for tighter quality control,

- more interfaces and hence higher potential for loss of reliability.

Figure 10.13 shows an example of differential product structure where, for a low production volume, it was less expensive to use this product structure than an integral product structure.

Integral product structures

Integral product structures combine several parts into a single part. Figure 10.14 shows an example where a cast and welded component has been replaced by a single cast component. Though the casting is fairly complex it led to a cost reduction of 36 per cent. The advantages and disadvantages of an integral product structure are easily determined by reversing the advantages and disadvantages of differential product structures.

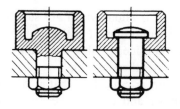

Figure 10.13 Differential product structure

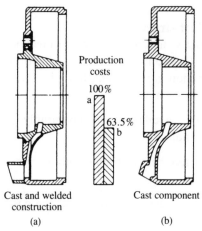

Figure 10.14 Integral product structure (a) cast and welded construction, (b) cast component

Further examples of differential and integral product structures with the appropriate conditions for each to occur are given in Figs. 10.15 to 10.22 (see pages 161 to 166).

Most real products will include examples of both differential and integral product structures and as such they are complementary. While, from the viewpoint of assembly, integral product structures are preferred, good judgement is needed to ensure that the requirements for assembly do not override valid requirements of other aspects of manufacture.

The rationalization of product structure can be approached in several ways. Manipulating the design degrees of freedom at the conceptual design stage leads to consideration of the following structures.

Integration	Differentiation
During operation of the product, does the part move relative to all other parts already assembled?	
No	Yes

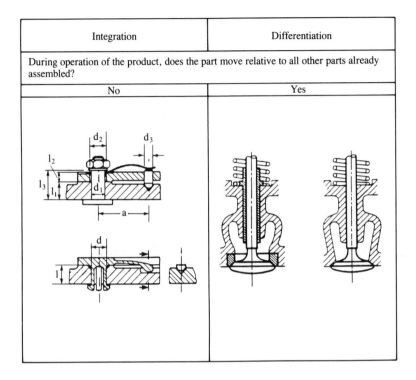

Figure 10.15 Example of integrated and differentiated assembly

Integration	Differentiation
Does the function realization depend on particular material characteristics?	
No	Yes

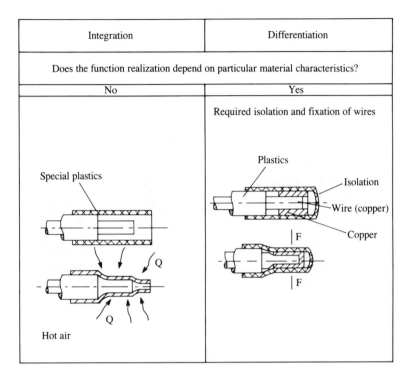

Figure 10.16 Example of integrated and differentiated assembly

Integration	Differentiation
Will some parts be easier to produce by division into sub-parts?	
No	Yes
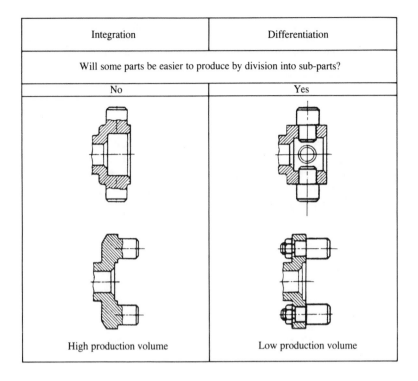	
High production volume	Low production volume

Figure 10.17 Example of integrated and differentiated assembly

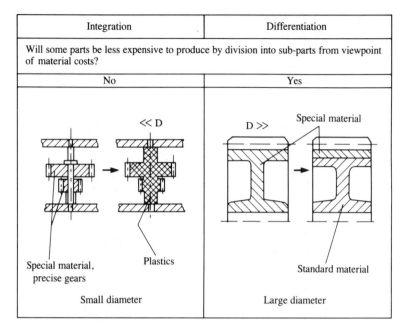

Figure 10.18 Example of integrated and differentiated assembly

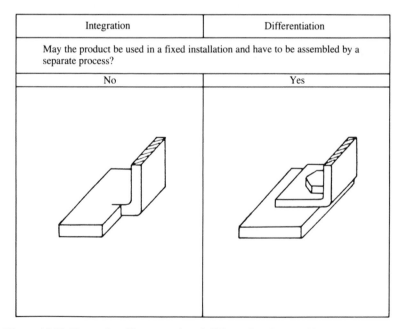

Figure 10.19 Example of integrated and differentiated assembly

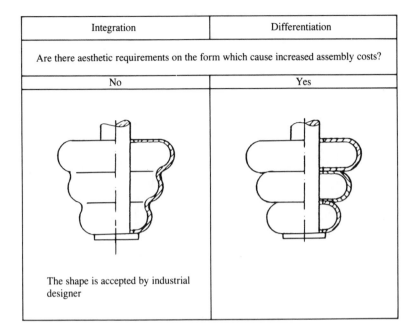

Figure 10.20 Example of integrated and differentiated assembly

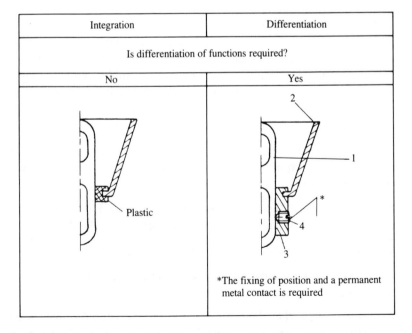

Figure 10.21 Example of integrated and differentiated assembly

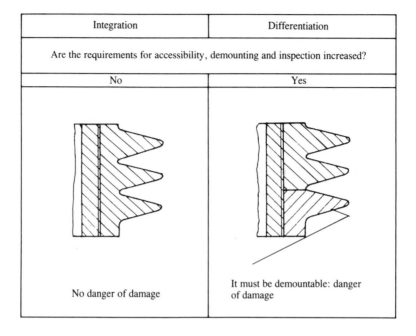

Integration	Differentiation
Are the requirements for accessibility, demounting and inspection increased?	
No	Yes
No danger of damage	It must be demountable: danger of damage

Figure 10.22 Example of integrated and differentiated assembly

Serial product structures

In a serial product structure, parts are arranged one over the other and there is little possibility for an alternative assembly sequence. Such a product structure is 'layered' or 'additive' assembly. Serial product structures come in two forms, those with different (Fig. 10.23) and those with the same (Fig. 10.24) parts, i.e. in shape and size.

A serial product structure has some advantages, such as:

- The assembly motions are straight, usually mutually perpendicular lines and the requirement for degrees of freedom of assembly equipment is reduced.

- Most of the assembly operations can be performed from the same direction and this would be chosen to be from vertically above.

- The stacked construction tends to reduce the number of fasteners.

- If the same parts are used, the number of insertion operations will tend to be reduced.

As pointed out earlier, however, a serial product structure will often have virtually no alternative assembly sequences and this is often an

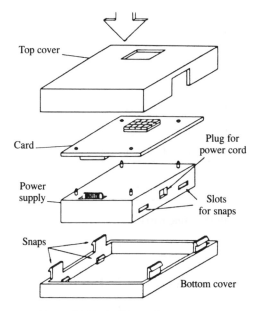

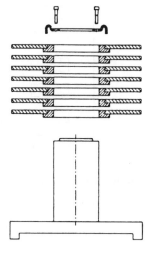

Figure 10.23 Serial product structure with different parts

Figure 10.24 Serial product structure with the same parts

undesirable characteristic. Of course, the potential for reduced number of parts might outweigh the loss of flexibility in sequencing.

Parallel product structures

A parallel product structure is an arrangement of parts such that parallel (simultaneous) assembly operations are possible. Parallel product structures can have the same or different parts, size and shape. The printed circuit board offers excellent examples of both types. Figure 10.25 shows a schematic of a board with similar part types although these parts could have different functionality. Figure 10.26 shows a more typical board which has parts that are different from the viewpoint of assembly. A parallel product structure has some advantages such as:

- Assembly motions are often direct vertical lines.

- The order of assembly operations is not fixed.

- There is usually good accessibility.

While there are some advantages to a serial product structure, in general, the significant sequencing advantages of a parallel product structure are usually more important than any benefits obtained from a serial product

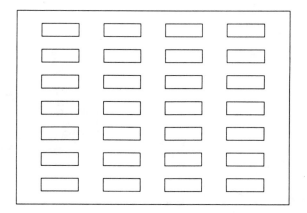

Figure 10.25 Parallel product structure with the same parts

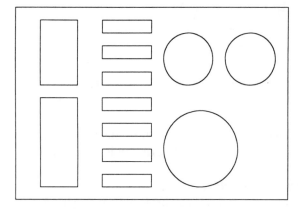

Figure 10.26 Parallel product structure with different parts

structure. With thought, many products can combine the best features of each type of product structure.

Combined product structures (building blocks)

If it is possible to divide a product into independently structured parts that are functional units with simple relationships between them, the product is said to be capable of being assembled from building blocks.

The advantages of such a product structure are:

• Building blocks can be 'changed' and tested independently.

• Assembly of the blocks can be carried out in parallel.

• Repairing of malfunctioning parts is easier.

Figure 10.27(a) shows a reduction valve placed directly in the body of a

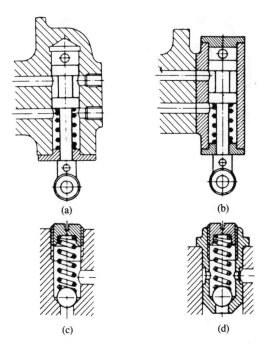

(a)   (b)

(c)   (d)

Figure 10.27 Building blocks

case. A precise slot is manufactured directly into the case and the performance of the valve can be diminished by presence of casting defects. The high demands on surface quality and tolerancing can be reduced by designing the valve as building blocks (Fig. 10.27(b)). Production is simplified, assembly is parallel and testing is easy. A further example is shown in Figs. 10.27(c) and 10.27(d) where a safety valve has been redesigned using building blocks.

Generally, it can be said that good product design would include an analysis which would consider the possibilities for dividing the product into building blocks. The material, manufacturing and assembly costs of the extra parts required are often more than offset by the considerations outlined above.

Product structures with good accessibility

The spatial arrangement of parts and their shape determines the accessibility to particular parts, fasteners, surfaces, inspection points, etc. and it is sensible to leave the product as 'open' as possible, particularly in strategically important locations. Parts with unavoidably low reliability, those that are easily damaged and those that need

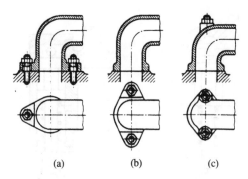

(a) (b) (c)

Figure 10.28 Good accessibility

inspection must be easily accessible. Figure 10.28(a) shows an example of a pipe construction with bad accessibility; better solutions are shown in Figs. 10.28(b) and 10.28(c).

Ensuring accessibility in important locations is very important.

Product structures with a base part

In a product structure a key element is the base part. It is the first part to appear in the assembly and it is usually the part into which all other parts fit (more parts are related to the base part than to any other). The existence of a unique base part in a product structure has some advantages such as:

- It can act as a fixture.

- It can eliminate force effects of joining process.

- It is often possible to integrate fasteners into the base.

Having ensured that a unique base part is embodied in the design, effort should be made that it is:

- stable

- there is good accessibility from all directions

- there are good grasping surfaces

Figure 10.29 illustrates these points.

Composite product structures

By manipulation of the number and arrangement of parts and their form, dimensions and material characteristics, a composite product structure

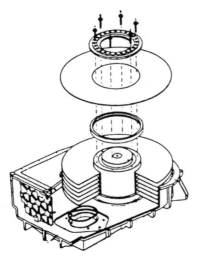

Figure 10.29 Base part

can be identified. A composite product structure is one where there is:

- the inseparable connection of several parts into a single component,

- the simultaneous application of several joining methods for the combination of parts,

- the combination of several materials for optimum exploitation of their properties.

Figure 10.30 shows a good example of composite construction. The tape drive chassis is produced by injection moulding and precise individual parts are needed. Using 'Hostaform', an injection moulding is produced with integral side plates in one operation. All the parts are fixed and are fully operational immediately; no finishing is required.

Total chassis

Very often, the base part of an assembly is itself a complex sub-assembly which apart from requiring assembly also results in tolerance build-up between the various elements of the base part. This has no importance in manual assembly, but in automated assembly, several datum points will require complex and expensive 'measuring' equipment to make assembly possible. A total chassis is a construction built from punched and bent steel (or plastic) plates, joined together into a highly precise part. Using a total chassis avoids 'fitting' and provides a good base part for assembly operations, assembly transfer and inspection.

Fig. 10.31 shows a simple feature of a modular chassis where rather

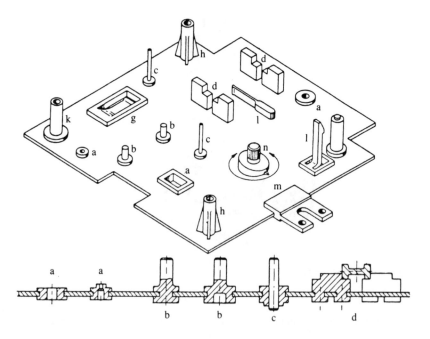

Figure 10.30 Compound construction

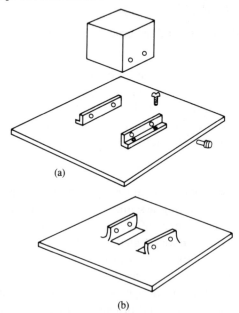

Figure 10.31 Total chassis concepts (a) fabricated support brackets, (b) punched and bent support brackets

than fabricate location features for attaching a component using angles and screws, the material is punched and bent to produce the same effect. Not only is the alignment of the holes better, the assembly and manufacturing costs have been reduced.

Assembly directions

From the point of view of assembly, it is best if the number of assembly directions is minimal. Assembly directions different from vertical require more degrees of freedom of the assembly equipment, or the turning of parts/the partial assembly or both. Assembly directions other than from vertically above are difficult for all assembly methods.

A designer by good selection of types of fastener and placement of parts determines the requirements for assembly directions. Figure 10.32 illustrates possibilities for reducing the number of assembly directions.

Assembly motions

The shape and arrangement of parts can cause complex assembly motions. Complex assembly motions require more degrees of freedom for assembly robots or more expensive dedicated equipment and the use of simple motions is advisable. Assembly motions (see Fig. 10.33) have considerable kinematic complexity; it is suggested that preferred

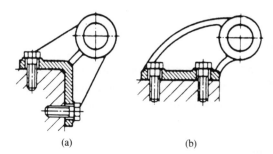

(a)   (b)

Figure 10.32 The minimization of assembly directions

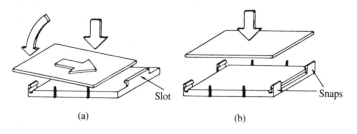

(a)   (b)

Figure 10.33 Compound and simple assembly motions

assembly motions are either uni-axial translation or uni-axial translation followed by simple rotation. Figure 10.33 shows how to simplify the complexity of assembly motions.

10.8 Joints, connections and fasteners

Joints

The individual rigid bodies which collectively form a mechanism are said to be members. Members may consist of non-rigid bodies, such as cables, liquid gas or formless material, which momentarily serve the same function as rigid bodies and are sometimes referred to as resistant bodies. The members are interconnected in pairs at points of contact called joints.

The part of a member's surface which contacts another member is called a pair element. The combination of two such elements constitutes a kinematic pair or connection. The connection is a combination of pair elements of members permanently kept in contact so that there exists or does not exist relative motion between them. Mobility of connection is defined by a number of degrees of freedom (NDF), according to which it is:

- movable (NDF is 1, 2, 3, 4, 5)

- solid (NDF $\leqslant$ 0)

It is important to differentiate a joint and joining techniques. A joint is a design feature necessary for functionality, joining is a process or processes characterized by the need for separate parts and the 'quality' of the force between the mating parts.

Movable connections

Movable connections are a type of permanent connection of parts in which relative motions are possible if working forces are acting. The mobility of movable connections is calculated in mechanisms theory by the number of degrees of freedom. A model of two rigid bodies with point contacts is used for this. The measure of the suitability of movable joints for assembly is dependent on the number of constraints of connection. A high number of constraints of connection requires a wide tolerance or additional adjusting operations.

Figure 10.34 shows an ideal single-degree-of-freedom movable joint, a real severely over-constrained joint and a better solution. Mostly, real

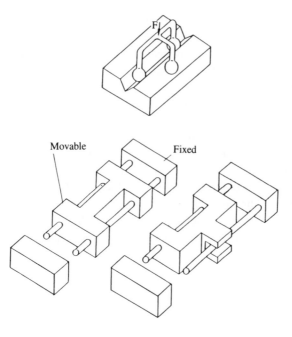

Figure 10.34 An ideal, a severely over-constrained and a less constrained construction

movable joints have too many constraints. This helps to increase the strength of a construction but at the expense of ease of assembly.

Fasteners

By definition, all fasteners are potentially redundant since their only function is to fasten. Progressively the best fasteners in order from best to worst are:

- Integral fasteners—these have no extra parts.

- Snap-fit fasteners, circlips, spire nuts, etc.—these only require elastic deformation.

- Deforming fasteners, rivets, knock-down tabs, etc.—these require plastic deformation.

- Threaded fasteners, nuts, bolts, etc.—significant time is needed to perform the operation.

- Ancillaries, washers, split pins, etc.—often the part can be eliminated.

Obviously, when functionality requires significant forces between fastened elements, the choice of suitable fastener is severely restricted. However, the function of many fasteners is to maintain relative positions and for this, too many fasteners of the wrong type are often used.

Each additional fastener increased the need for magazines, feeders, tools, etc. Figure 10.35 is a simple example illustrating how to reduce the number of separate fasteners. Several self-tapping screws do not need pre-drilled holes. They are normally used on light-gauge metal for such things as sheet-metal ducts. Self-piercing screws have a pointed tip that forms a pilot hole under pressure from the assembly tool. After the point is driven into material, the threaded portion of the fastener forms threads as it is turned. The metal deformation caused by forming the pilot hole increased the thread-engagement area. Self-drilling screws have a self-contained bit that drills a hole. Once the metal is penetrated, the fastener functions as a conventional self-tapping screw.

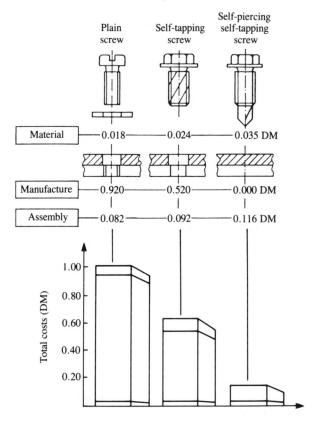

Figure 10.35 The total costs of the different types of screw joints

For this type of parts reduction, care has to be taken to ensure that the benefits in assembly are greater than the costs in parts.

Special fasteners

Special fasteners offer mechanical and functional properties which standard ones do not have. A wide variety of special-purpose fasteners are available for applications where standard fasteners are inappropriate or insufficient. Usually, special materials, such as spring steel, hot or cold formed metal/plastics are used for fasteners such as clips, clamps and cable ties. Self-sealing fasteners can be used to help keep gases and liquids in or out of an assembly. Quick-operating fasteners are used in applications where quick or repeated access to a part is necessary.

Spring clips are self-retaining, one-piece fasteners that slip into a mounting hole or onto a flange or panel edge. Secondary fastening devices such as rivets, studs, or screws are not necessary because spring clips are held by spring tension and do not loosen easily through vibration (Fig. 10.36).

Cage nuts are particularly useful in blind fastening locations. Their self-retaining feature eliminates the need for welding, clinching or staking the nut in place. They can be installed after painting or coating, making masking or re-tapping unnecessary (Fig. 10.37).

Quick-operating fasteners

These fasteners should be used when repeated access to a part is necessary. When selecting these fasteners, important factors include strength of construction, smoothness of operation and easy installation. Some types of camlock quick-operating fasteners are shown in Fig. 10.38.

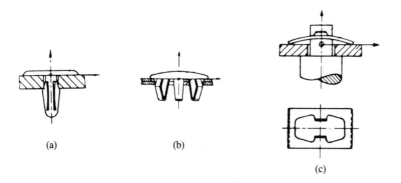

(a)   (b)   (c)

Figure 10.36 Examples of spring clips

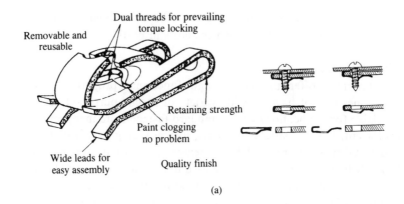

(a)

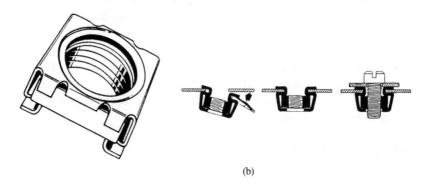

(b)

Figure 10.37 Examples of cage nuts

Cable and tube clips

Retaining elements to hold wires or tubing are incorporated into these
fasteners which engage panel holes, mounting flanges, or panel edges.
The fasteners are front mounting and require no access to the back of the
panel. One spring clip can replace several loose parts. Figure 10.39 shows
several examples of cable and tube clips.

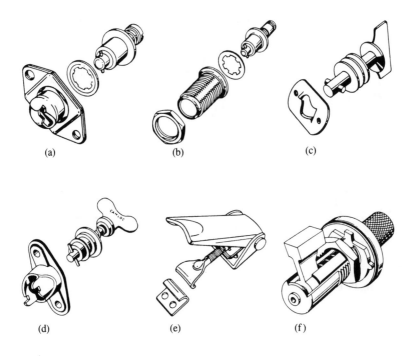

(a)   (b)   (c)

(d)   (e)   (f)

Figure 10.38 Examples of quick-operating fasteners

10.9 Part stability

The stability of a part's location is important especially if the partial assembly is transported during assembly. Parts assembled vertically and for which the sub-assembly does not rotate about a horizontal axis only need lateral constraint (Fig. 10.40). Parts assembled horizontally and maintained in this position, or parts which are turned such that lateral constraint is removed, need 'semi-fixing' (Fig. 10.41). The alternative in automated assembly is very expensive 'holding down'.

10.10 Unambiguous insertion

Perhaps the most difficult assembly operation to perform, even by people, is insertion where two separate sites contact simultaneously. An example of this is shown in Fig. 10.42 together with an appropriate solution.

Flexible conduit clip   One-piece clamp   Plastic wire retainer

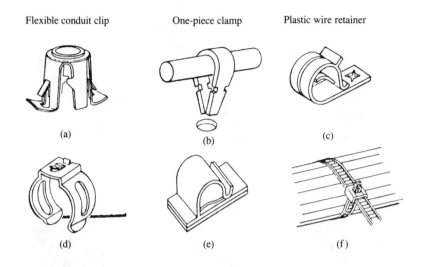

(a)   (b)   (c)

(d)   (e)   (f)

Figure 10.39 Cable and tube clips

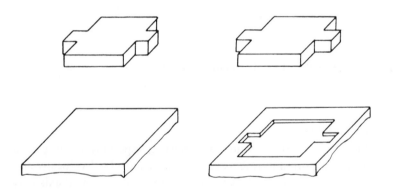

Figure 10.40 Securing of placement for vertically assembled components

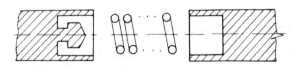

Figure 10.41 Securing of placement for horizontally assembled components

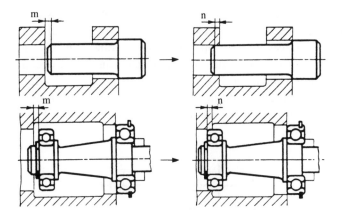

Figure 10.42 Redesign to eliminate simultaneous insertion of components

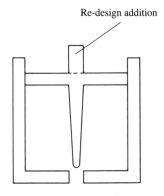

Figure 10.43 Redesign to improve the quality of the grasp just prior to the release of the part

10.11 Unattainable insertion

This occurs when the grip site cannot be maintained up to the point of initial insertion. An example of this and a possible solution is shown in Fig. 10.43.

10.12 Assembly surfaces

Placement

It is sensible to avoid simultaneous assembly involving several assembly surfaces. It is also bad practice to have assembly surfaces crossing. Fig. 10.44 illustrates how to avoid assembly having several simultaneous assembly surfaces with an open positive tolerance (in this case a gap), where the gap accommodates deviations in production. Crossed assembly surfaces are shown in Fig. 10.45 (a), (b), (c), and (d); all these constructions require additional machining of assembly surfaces after assembly of the top cover. Better solutions are shown in Fig. 10.46 (a), (b), (c), and (d).

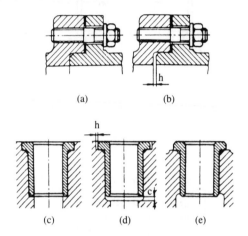

Figure 10.44 Avoidance of simultaneous assembly surfaces

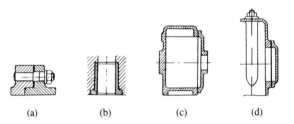

Figure 10.45 Potential for crossed assembly surfaces

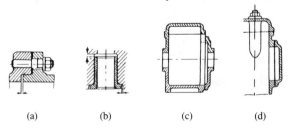

Figure 10.46 Designs to eliminate crossed assembly surfaces

Shape and precision

Assembling is easy if assembly surfaces are plane. Contact points on cylindrical surfaces should generally be avoided because assembly is usually less positive and, consequently, more difficult and less reliable. Precise assembly surfaces should be minimized. If large precise surfaces are required, the result will be additional production costs. Fig. 10.47 demonstrates these principles.

10.13 'Obvious assembly'

The arrangement of parts and their shapes must be such that the placing of them is unambiguous. Subjective judgements in assembly are usually undesirable. Where, and how, parts are positioned should be clear to an operator. Fig. 10.48(a) illustrates this principle schematically. An example of ambiguous assembly is shown in Fig. 10.48(b).

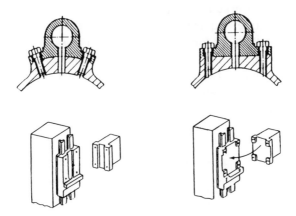

Figure 10.47 Elimination of precise assembly surfaces

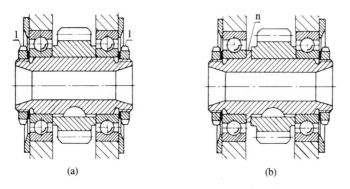

Figure 10.48 Design to ensure that assembly is (a) 'obvious', (b) 'ambiguous'

10.14 Tolerancing

Precise placing of parts on assembly surfaces in relation to each other implies additional manufacturing and assembly cost or 'fitting'. This can be avoided by the use of 'elastic elements' or a reduction in tolerance builds-up by a reduction in the number of parts. Figure 10.49(a) shows an example of the former, Fig. 10.49(b), an example of the latter.

A further common tolerance design fault is to have open tolerancing on gripped dimensions relative to insertion dimensions. Clearly in manual assembly this does not matter but in automated assembly it can cause significant insertion difficulties.

10.15 Dealing with variants

Variant-oriented product design

The current market requires products with variants and these requirements increase the need for flexibility of assembly systems. The question is, how to design products with variants that minimize the need for extra flexibility. The link between flexibility requirements and variants can be defined as variance, where variance is the degree of variation that can be built into a product to reduce the assembly problems caused by variants.

The influence of the designer in the two tasks of designing the product

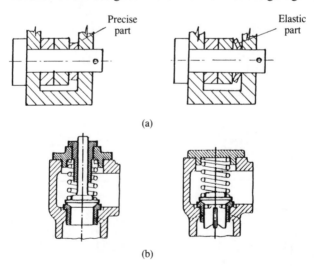

(a)

(b)

Figure 10.49 (a) Good design with one elastic element, (b) reduced tolerance build-up by having less parts

and deciding on the assembly system can be considered to require:

- not making demands for flexibility in the assembly equipment when designing the product,

- exploitation of any flexibility that assembly equipment might inherently possess when designing the product,

- designing the product so as to obtain new, increased flexibility in assembly.

The general principles which can be used to help design for flexibility with respect to the first two points above will be discussed later in the chapter.

Starting from the situation where one or more products are to be assembled in an assembly system and one or more products need to be added to this group, when designing the new products, the crucial question which the designer needs to address with respect to assembly is: 'How can the products be designed so that they do not make any new demands on the flexibility of the assembly equipment?' For a designer, it is important to know how to design the product so as to make product variance transparent to the assembly system. There are three basic principles for the design of products with variants:

- avoid variants,

- standardize at all levels of product design,

- create modular designs.

These principles will now be discussed. In the main this is with reference to automated assembly and, in particular, dedicated assembly. However, while manual assembly can more easily accommodate variants than any form of automated assembly, the general principles for dealing with variants apply to all methods of assembly.

Avoidance
It must be assumed that a 'request' by the marketing/sales department for a new variant is well founded since the consequences for assembly costs are potentially severe. Before a new variant is introduced, all the implications and possibilities need to be considered. Perhaps a slightly modified existing variant could be used; perhaps the design of the new variant could be tailored to suit the capabilities of the existing assembly equipment; perhaps the assembly equipment needs to be evaluated in the light of current and future requirements.

Product flexibility allows a company to be responsive to the market by enabling it to bring newly designed products quickly to the market. Since future product designs are usually unknown, it becomes important to not only design and develop 'production-friendly' products, but also to continuously review and evolve appropriate production equipment.

Standardization

The product variance is the degree of variation which is built into the product and which determines how many variants can be made. One effective method of improving variant-oriented production is standardization. Fig. 10.50 shows a variant-oriented product produced by Nippondenso Co. Ltd. Here, by designing the indicator instruments appropriately, it was possible to standardize the individual components such that 17 different parts can now produce 288 variants, whereas before, 48 different parts could only produce 150 variants.

Modularity

The second important principle of rationalization for variant-oriented production is modularity. Modularity implies the product can be built as a series of building blocks where all the product variants can be created with sets of self-contained parts having standard interfaces to other parts or sub-assemblies in the system. Modular design offers the ability to standardize diversity because it allows a product to be customized by using different combinations of building blocks.

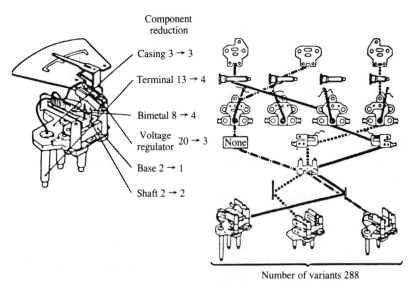

Component
reduction

Casing 3 → 3

Terminal 13 → 4

Bimetal 8 → 4

Voltage
regulator 20 → 3

Base 2 → 1

Shaft 2 → 2

None

Number of variants 288

Figure 10.50 Advantage of standardization

As a general rule, it is useful to analyse the possibilities for using modular design.

A modularly structured product resists obsolescence and shortens the redesign cycle. A new generation of product can usually utilize most of the old modules, costs are reduced and ease of service and repair are enhanced because a non-functional module can be quickly replaced by a good one.

Excellent examples of modular design can be found in manufacturing industry where 'interfaces' have been developed to match machines to tools. A good and simple example of this is the morse taper sleeve (Fig. 10.51) where a range of sleeves allow a wide variety of drills to fit in a wide variety of machine spindles. Another good example is modular fixturing where a set of generic fixturing elements can be used in many combinations to meet different fixturing requirements. Modularity often increases assembly complexity but it increases flexibility and usually reduces manufacturing and replacement costs.

Number of sub-assembly variants

If a new product variant is needed, this should use as few sub-assembly variants as is possible since the fewer the total number of sub-assembly variants, the more use can be made of common building blocks. Figure 10.52 illustrates this rule.

Time at which variants are manufactured

Variant features should appear as far up the assembly chain as is possible.

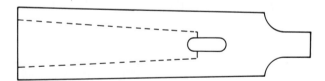

Figure 10.51 Rationalization using modularity

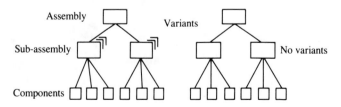

Figure 10.52 Minimizing the number of sub-assembly variants

For effective assembly of product variants, the majority of assembly needs to be common to all variants. Figure 10.53 illustrates toys (Lego) for which the body and arms of animals are common components in all variants. The variants are produced at the end of the assembly process where different coloured and different shaped heads are used. For added variety, the bodies are also produced in different colours but, of course, this variance is transparent to the assembly equipment.

Base part
If a new product variant is needed, do not change the base component since this almost inevitably would result in the need for a new fixture. An example using existing components is shown in Fig. 10.54.

Extra connections
Extra connections between variant building blocks and non-variant building blocks require extra assembly operations and this should be avoided. An example of a requirement and good and bad solutions is shown in Fig. 10.55.

Assembly directions
Uni-directional assembly is clearly most appropriate since it produces lowest cost assembly. Many real products cannot meet this condition and repositioning of the partial assembly becomes inevitable. Producing new

Figure 10.53 Making variants as late as possible

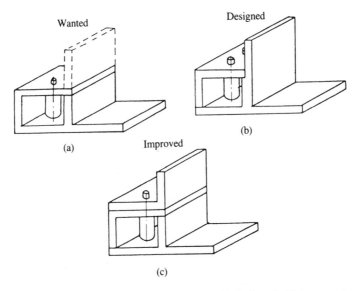

Figure 10.54 The base component (a) wanted, (b) designed, (c) improved

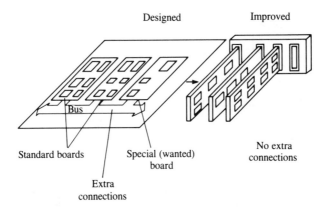

Figure 10.55 Extra connections

variants with an assembly direction not needed beforehand will almost always result in increased assembly costs and an extra, already-used, assembly direction will often increase costs. Figure 10.56 illustrates an example of a bad design change.

Assembly positions
A new position of a part in an assembly will require additional programming and might require additional equipment. In the worst case, complete reprogramming could be needed and most of the

equipment would need replacing. Figure 10.57 illustrates an example of a bad design change.

Assembly sequence

Never introduce a variant which changes the assembly sequence. At least, expensive rearrangement of equipment would be necessary; in some cases equipment would need to be replaced. Figure 10.58 illustrates an example of a bad design change.

Standardization of parts

If a new product variant requires extra parts, ensure as far as is possible that the extra parts are ones already used for different applications, particularly in the same assembly. For dedicated assembly this approach results in further use of existing technology (no development or commissioning costs); for robot assembly it results in fewer tools and fewer parts feeders. For all forms of assembly, fewer, larger quantities of parts are needed and this has both cost of procurement and cost of

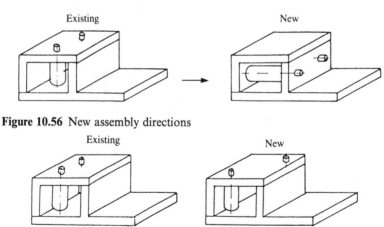

Existing   New

Figure 10.56 New assembly directions

Existing   New

Figure 10.57 New assembly positions

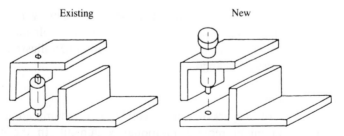

Existing   New

Figure 10.58 New assembly sequence caused by new variant

administration implications. Figure 10.59 illustrates two examples showing bad and improved design.

Standardization of grasp and assembly surfaces

If a new product variant requires the shape of the component to change, try to ensure that the grasp and assembly surfaces remain unchanged. As far as the equipment is concerned, the assembly task is unchanged. Figure 10.60 shows an example in which the original design had three different grasp surfaces for the parts involved, and where, for the new design, there is only one.

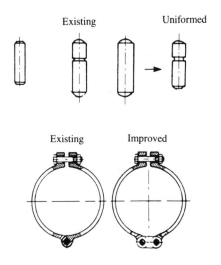

Figure 10.59 The standardization of type of components

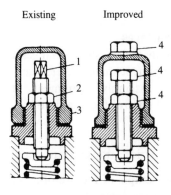

Figure 10.60 The standardization of grasping surfaces

Standardization of feeding and orienting features

If a new product variant requires the shape of the component to be changed, try to keep the features used for orientation in a small parts feeder the same. This will ensure that a new small parts feeder is not required. Figure 10.61 shows an existing variant, the required change and a design which embodies the change without changing the feeding and orienting features. Figure 10.62 shows an example of an existing orienting feature, the change required and a solution that allows orientation without the need for a new parts feeder.

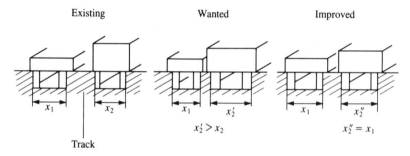

Existing   Wanted   Improved

$x_2' > x_2$ $x_2'' = x_1$

Track

Figure 10.61 The same feeding features

Existing   Wanted   Improved

Figure 10.62 The same orienting features

Quality of assembly surfaces

Experience has shown that even small changes in tolerances, surface quality, deburring, component material, etc. can cause disruption and hence modification to assembly equipment. It is, therefore, very important to specify and control the quality of:

• fixing surfaces,

• surfaces used for orientation, transportation and insertion.

Figure 10.63 shows an example of the use of a better alignment feature which will help in facilitating assembly.

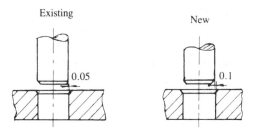

Figure 10.63 Too small tolerances are a problem in assembly

10.16 Conclusion

The purpose of this book has been to show product designers the parameters that are important when considering the assemblability of a product. Care has been taken to ensure that DFA has not been overemphasized to the extent that all the other important elements of product design are ignored. Nevertheless, with assembly accounting for a large proportion of manufacturing costs, and particularly in a world environment where there is a large disparity between assembly costs from nation to nation, it is very important that those with high unit labour costs control product costs by reducing the assembly content of work.

It is hoped that the book has promoted some of the rather less publicized factors which significantly affect assembly and that it has indicated the relative importance of the many, sometimes contradictory factors. It must be remembered that, of the multitude of factors which influence assembly and of the even larger number of conflicts which exist between design for assembly and design for manufacturability, functionality, appearance etc., there are few important factors and few important conflicts.

To remember everything contained in the book presents a problem. However, many of the suggestions for good design for assembly are second-order effects and care should be taken not to place too much emphasis on them. In product design at the conceptual stage the priorities must be:

- Establish the minimum number of parts—by far the most important objective. If only this were done then, for any method of assembly, assembly costs will be competitive.

- Minimize non-productive operations—for most products, there are too many examples of costs being incurred without any contribution being made to assembly. In automated assembly this is particularly true.

- Create 'open' precedences—virtually no attention is given to assembly sequence apart from considering the problems of line balance. Since balance by definition is only associated with assembly lines and these account for an almost insignificant proportion of assembly, it is foolish not to consider the wider aspects of sequencing and to design to give more sequencing opportunities.

With these very basic methodologies, and lateral thinking, much can be done to create not only good designs for assembly but also better, less costly products.

Bibliography

Allen, A. J., Bielby, M. S., Swift, K. G., Development of a Product Manufacturing Analysis and Costing System, *Int. J. Adv. Manuf. Technol.*, vol. 6, pp. 205–215, 1991.

Ambler, A. P., Popplestone, R. J., Inferring the Positions of Bodies from Specified Spatial Relations, *Artif. Intelligence*, vol. 6, pp. 137–174, 1975.

Andreasen, M. M., Kahler, S., Lund, T., *Design for Assembly*, Springer-Verlag, New York, 1983.

—Hein, L. *Integrated Product Development*, IFS (Publications), Springer-Verlag, New York, 1985.

Andresen, U. Ein Beitrag zum methodischen Konstruktieren bei der montage-gerechten Gestaltung von Teilen der Grosserienfertingung, Diss, TU Braunschveig, no. 9, 1973.

Angemuller, G., Moritzen, K., A Knowledge-Based System Supporting Product Design for Mechanical Assembly, *Proc. 1st Conf. AI and ES in Manuf.*, London, pp. 181–192, March 1990.

Bassler, R., *Integration der Montagegerechten Produktgestaltung im den Konstruktionsprocess*, Diss. Univ., Stuttgart, 1988.

—Rationalisation Results in Assembly-Oriented Design, *Assem. Automation*, vol. 8, no. 2, pp. 82–86, 1988.

Boothroyd, G., Dewhurst, P., *Design for Assembly Handbook*, Univ. of Mass., Amherst, MA, 1983.

—*Product Design for Assembly Handbook*, Wakefield, RI, 1987.

—Product Design for Manufacture and Assembly, *ME*, April, pp. 42–46, 1988a.

—Estimate Costs at an Early Stage, *AM*, Aug, pp. 54–57, 1988b.

—New Developments in DFMA, *Proc. 6th Int. Conf. DFMA '91*, Newport, RI, 1991.

—Knight, W. A., Research Program on the Selection of Materials and Processes for Component Parts, *Int. J. Adv. Manuf. Technology*, vol. 6, pp. 98 ff, 1991.

—Radovanovic, P., Estimating the Cost of Machined Components during the Conceptual Design of a Product, *CIRP Annals*, vol. 38, no. 1, pp. 157–160, 1989.

Chal, J., *Methodology of Evaluation and Improvement of Assemblability of Product*, PhD thesis, Slov. Techn. University, Bratislava, CS, 1992.

—Redford, A., *Product Design for Manufacture and Assembly*, Techn. Report, Univ. of Salford, 1989.

Chay, D., Lenz, E., Sphitalni, M., Picking the Parameters to Easy Automation of Assembly, *Assemb. Automation*, vol. 8, no. 3, pp. 151– 154, 1988.

Clark, K. B., Fujimoto, T., Overlapping Problem Solving in Product Development. In *Managing International Manufacturing*, Ferdows, K. (ed.), Elsevier Science Publishers BV (North-Holland), 1989.

Davidson, R., Redford, A., *Product and System Design for Robot Assembly*, Robotic Assembly Consultants, Southport, UK, 1990.

Dewhurst, P., Boothroyd, G., Cutting Assembly Costs with Molded Parts, *MD*, vol. 21, pp. 68–72, 1988.

—Kuppurajan, D., Determination of Optimum Processing Conditions for Injection Molding, *Int. J. Prod. Res.*, vol. 27, no. 1, pp. 21–29, 1989.

Dick, M. D., *Computer Aided Design and Manufacturing System for Parts Feeding in Robot-Based Assembly*, PhD. thesis, Univ. of Salford, UK, 1989.

Dilling, H. -J., *Methodisches Rationalisieren vor Fertigungsprozessen an Beispiel Montagegerechten Gestaltung*, Diss. Techn. Hochschule, Darmstadt, 1978.

Ehrlenspiel, K. Genauigkeit, Gultigkeitsgrenzen, Aktualisierung der Erkemnutnisse und Hilfsmittel zum kostengunstigen Konstruieren, *Konstruktion* 32, 12, pp. 487–492, 1980.

—*Kostengustig Konstruieren*, Springer-Verlag, 1985.

Fujimoto, T., *Organisation for Effective Product Development: The Case of the Global Automobile Industry*, PhD. thesis, Harvard University, Graduate School of Business Administration, Boston, 1989.

Gairola, A., *Massnahmenkatalog fur das Montagerechte Konstruieren*, VDI-Ber. 596, VDI-Verlag, Dusseldorf, 1987.

Hager, C. D., Kim, S. H., Information and its Effect on the Performance of a Robotic Assembly Process, *Proc. Symp. on Intelligent and Integrated Manufacturing: Analysis and Synthesis*, ASME Winter Annual Meeting, Boston, pp. 349–356, 1987.

Hitachi, Ltd., Quantitative Evaluation of Assemblability by Design Drawing, *Nikkei Mechanical*, Oct. 2, 54–55, 1978.

Henderson, M., Feature Definition Techniques in Automated Analysis. In *Geometric Modeling for Product Modeling*, Wozny, M. J., Turner, J. U., Preiss, K. (eds), Elsevier Science Publishers BV (North-Holland), IFIP, pp. 321–335, 1990.

Hernani, J., *An Expert System Approach to the Choice of Design Rules for Automated Assembly*, MSc. thesis, Cranfield Institute of Techn., Bedford, UK, 1985/86.

Holbrook, A., Sackett, P., DFA as a Primary Process Decreases Design Efficiencies, *Assem. Automation*, Aug., vol. 8, no. 3, 137–140, 1988.

—Design for Assembly through Knowledge Application, *Assem. Automation*, May, pp. 87–92, 1990.

Homen de Mello, L. S., *Task Sequence Planning for Robotic Assembly*, PhD. Thesis, Carnegie-Mellon Univ., Pittsburgh, PA, 1989.

—Sanderson, L. A. C., Two Criteria for the Selection of Assembly Plans: Maximizing the Flexibility of Sequencing the Assembly Tasks and Minimizing the Assembly Time through Parallel Execution of Assembly Tasks, *IEEE Trans. on Robotics & Automation*, vol. 7, no. 5, pp. 266–633, 1991.

Imai, K., Nomaka, I., Takeuchi, H., Managing the New Product Development Process: How Japanese Companies Learn and Unlearn. In *The Uneasy Alliance: Managing the Productivity Technology Dilemma*, Clark, K. B., Hayes, R. H., and Lorenz, C. (eds), Harvard Business School Press, Boston, pp. 337–375, 1985.

Jackson, D. H., *Effects on Product Design of Components Assembled by Automatic/Robotic Assembly*, MSc. Thesis, Cranfield Ins. of Techn., Bedford, UK, 1985.

Joshi, S., *CAD Interface for Automated Process Planning*, PhD Thesis, School of Industrial Engineering, Purdue Univ., W. Lafayette, IN, 1987.

—Feature Recognition and Geometric Reasoning for some Process Planning Activities. In *Geometric Modeling for Product Engineering*, Wozny, M. J., Turner, J. U., Preiss, K. (eds), Elsevier Science Publishers BV (North-Holland), IFIP, pp. 363–384, 1990.

Kim, S. H., *Mathematical Foundation of Manufacturing Science: Theory and Implications*, PhD Thesis, MIT, 1985.

Kroll, E., Lenz, E., Wolberg, J. R., A Knowledge-Based Solution to the Design-for-Assembly Problem, *Manuf. Rev.* vol. 1, no. 2, June, pp. 104–108, 1988.

Langmoen, R., *Assembly with Robots*, PhD Thesis, Univ. Trondheim, Norway, 1984.

Lascz, J. F., Product Design for Robotic and Automatic Assembly. In *Robotic Assembly*, Rathmill, K. (ed.), IFS Publications, Ltd, Bedford, UK, pp. 157–172, 1985.

Libardi, E. C., Dixon, J. R., Simmons, M. K., Computer Environments for the Design of Mechanical Assemblies: Research Review, *Eng. with Computers*, pp. 121–136, 1988.

Lotter, B., Using the ABC Analysis in Design for Assembly, *Assem. Automation*, vol. 4, no. 2, pp. 80–86, 1984.

Maczska, K., General Electric has 'GE' method, *Design Eng.*, no. 2, pp. 148–151, 1982.

Marefat, M., Kashyap, R. L., Geometric Reasoning for Recognition of Three-Dimensional Object Features, *IEEE Trans. on Pattern Anal. Mach. Intelligence*, vol. 12, no. 10, Oct, pp. 949–965, 1990.

Michaels, J. V., Wood, W. P., *Design to Cost*, John Wiley & Sons, NY, 1989.

Miyakawa, S., Ohashi, T., The Hitachi Assemblability Method, *Proc. 1st Int. Conf. on Product Design for Assembly*, Newport, RI, 1986.

—Inoshita, S., Shigemura, T., The Hitachi Producibility Evaluation Method (PEM), *Kikai-sekkei, Nikkonkoyyo-shinbusha*, vol. 33, no. 7, pp. 39–47, 1989.

—Inoshita, S., Shigemura, T., The Hitachi Assemblability Evaluation Method (AEM), *Proc. Int. Conf. on Mfg. Systems and Environment—Looking Toward the 21st Century (The Japan Society of Mech. Engrs.)*, pp. 277–282, 1990.

—Iwata, M., The Hitachi Assemblability Evaluation Method (AEM), *Proc. 18th NAMRC Conference SME*, pp. 352–359, 1990.

Moritzen, K., Wissenbasierte Bewertung der Montagegerechtheit von Produktentwurfen. In *Expertensysteme in Entwicklung und Konstruktion*, VDI-Ber., no. 755, VDI-Verlag, Dusseldorf, pp. 235–250, 1989.

Nakazawa, H., Suh, N. P., Process Planning Based on Information Concept,

Robotics & CIM, vol. 1, no. 1/2, pp. 115–123, 1989.

Ong, N. S., Boothroyd, G., Assembly Times for Electrical Connections and Wire Harnesses, *Int. J. Adv. Technol*, no. 6, pp. 155–179, 1991.

Perry, T. S. (ed.), Teamwork plus technology cuts development time, *IEEE Spectrum*, vol. 27, no. 10, pp. 61–67, 1990.

Pham, D. T., Yeo, S. H., A Knowledge-Based System for Robot Gripper Selection Criteria for Choosing Grippers and Surfaces for Gripping, *Int. J. Mach. Tools Manuf.*, vol. 28, no. 4, 301–313, 1988.

Poli, C., Fenoglio, F., Designing Parts for Automatic Assembly, *MD*, vol. 10, pp. 79–84, 1987.

Popplestone, R., Liu, R. J., Weiss, R., A Group Theoretic Approach to Assembly Planning, *AI Magazine*, pp. 83–97, Spring 1990.

Popplestone, R. J., *et al.*, RAPT: A Language for Describing Assemblies, *The Industrial Robot*, vol. 5, no. 3, pp. 131–137, 1978.

—An Interpreter for a Language for Describing Assemblies, *Artif. Intelligence*, vol. 14, pp. 79–107, 1980.

Reacy, P. T., Ochs, J. B., Ozsoy, T. M., Wong, M., Automated Tolerance Analysis for Mechanical Assemblies Modeled with Geometric Features and Relational Data Structure, *CAD*, vol. 23, no. 6, 1991.

Rosario, L. M., Design for Assembly Analyses: Extraction of Geometric Features from a CAD System Data Base, *Annals of the CIRP*, vol. 38, no. 1, pp. 13–16, 1989.

Roy, U., Liu, C. R., Woo, T. C., Review of Dimensioning and Tolerancing Representation and Processing, *CAD*, vol. 23, no. 7, pp. 466–483, 1991.

Sherrin, I., Jared, G., Limage, M. G., Swift, K. G., Automated Manufacturability Evaluation, *Proc. 6th Int. Conf. On DFMA*, Wakefield, RI, 1991.

Shah, J. J., Assessment of Features Technology, *CAD*, vol. 23, no. 5, pp. 331–343, 1991.

Smith, P. G., Fast-Cycle Product Development, *Eng. Manuf. Journal*, vol. 2, no. 2, pp. 11–16, 1990.

Sphitalni, E. *et al.*, Automatic Assembly of Three-Dimensional Structures via Connectivity Graphs, *Annals of the CIRP*, vol. 38, no. 1, pp. 25–28, 1989.

Stinsson, T., Teamwork in Real Engineering, *MD*, vol. 22, pp. 99–104, 1990.

Stoll, H., Design for Manufacture, *ME*, pp. 67–73, 1988.

Subramani, P., Dewhurst, P., Automatic Generation of Product Disassembly Sequences, *Annals of the CIRP*, vol. 40, no. 1, pp. 115–118, 1991.

Suh, N. P., Orthonormal Processing of Metals, Part I Concept and Theory, *J. Eng. Ind.,* 104, pp. 327–331, 1982.

—Basic Concepts in Design for Producibility, *Annals of the CIRP*, vol. 37, no. 2, pp. 559–567, 1988(a).

—*The Principles of Design*, Oxford University Press, NY, 1988(b).

Swift, K. G., *Knowledge-Based Design for Manufacture*, Kogan Page, London, 1987.

—Production Oriented Design: A Knowledge-Based Approach, *Adv. Manuf. Eng.*, vol. 1, pp. 67–73, 1989.

Taguchi, G., *Introduction to Quality Engineering*, Asian Productivity Organisation, Tokyo, 1986.

—*System of Experimental Design*, ASI Press and Kraus International Publications, NY, vol. 1, 1987.

Takahashi, K., Senba, K., Design for Automatic Assembly, *Proc. 7th ICAA*, Zurich IFS Publications, Kempton, pp. 149–159, 1986.

Weissmantel, H., Regeln fur die Montagegerechte Produktgestaltung, *Autom. der Montage in der Feinwerktechnik und Elektrotechnik*, VDI-Ber. 747, VDI-Verlag, Dusseldorf, 1989.

Warnecke, H.-J. *et al.*, Montageautomatisierung begint bei der Produtkstaltung, *Konstruktion*, vol. 41, pp. 416–421, 1989.

Index